HAVE I GOT A STORY FOR YOU
Alex I. Askaroff
Copyright © Alex I. Askaroff

The rights of Alex I. Askaroff as author of this work have been asserted by him in accordance with the Copyright, Designs and Patents Act 1993.

All rights reserved. No part of this publication may be reproduced, stored in a retrieval system, or transmitted, in any way or form, or by any means, electronic, mechanical, photocopying, or otherwise, without the prior permission of the author and publisher.

British Library Cataloguing in Publication Data. A catalogue record for this book is available from the British Library.

BISAC Subject Headings: Biographical CIP Library Reference: Biographical

FOR OTHER TITLES BY THIS AUTHOR SEARCH AMAZON, GOOGLE, OR VISIT SEWALOT.COM

DEDICATION

My heart felt thanks go to my family and friends who add a little sweetness to the sometimes bitter pill of life, and to my customers who have filled my years with so many awesome tales.

ACKNOWLEDGEMENT

Front cover picture designed by Sarah Askaroff with kind permission of Katie Reilly. The picture is of her mum and dad, Sophie & Daniel Townsend hop-picking in Goudhurst in 1954. I had been looking for a front cover picture for two years until the day that Katie's sewing machine went wrong and I walked in, to see her parents staring down at me from the photo on the wall. I knew instantly that I had found it.

CONTENTS

AUTHOR'S NOTE

INTRODUCTION – THE SEWING WORLD IN EASTBOURNE

POEM – BIRTH IN A SEASIDE TOWN

HAVE I GOT A STORY FOR YOU

BABYLON WOODS

THE CAVENDISH POLTERGEIST

POEM – BROKEN DREAM

THE DAY THAT SUSSEX DIED

POEM – GOODNIGHT OLD MAN

WINTER CALLS

DAYS OF OUR LIVES

POEM – SIR MARMADUKE

THE REAPER COMETH

POEM – TINY ACORNS

CORAL

POEM – SUNLIGHT ON A SPECIAL DAY

POEM – THE RAINMAKER

BALLS

POEM – WHERE ARE YOU?

ISABEL

POEM – SPRINGTIME BLUES

THE RARY

POEM – THE CHRISTMAS GOOSE

TINA'S OLD NAPLES' PIZZA

HALF WAY TO HEAVEN

THE SUPPRESSOR MAN COMETH

POEM – CHILLY CHARLIE

THE GHOST OF SIR ARTHUR CONAN DOYLE

KERUNCH

WHO DARES WINS

EPILOGUE

AUTHOR'S NOTE

My stories build slowly. They are a labour of love. I learnt a lot from my first book, *Patches of Heaven*, and I now take a huge amount of trouble to try and make each page right. Firstly I make notes, taking down the story from my customer. I then type out a rough copy and research any discrepancies. I often phone or visit my customer again and again until I get everything just right. I then read the story aloud, correcting as I go, playing with the words like a jugglers plays with his clubs or balls, getting each sentence just right. Yana then works through the story. I do the alterations, and then my beautiful daughter-in-law, Corrinne, corrects more grammar. She is the best at it, so I leave her until last.

By now each story flows and when they are all done and dusted, I send them all to my publisher. They carry out the final edits and send me the book back for the last polish. If my book was a car it would have had at least 20 washes, a dozen polishes and endless waxes.

Hopefully all this effort shows in the following pages. What appears effortless has my blood flowing through each page. Every writer will know what I am talking about. They call it 'their struggle'.

INTRODUCTION
THE SEWING WORLD IN EASTBOURNE

You must agree that it is a strange old world. As I drive around the south east of England in my old banger of a Land Rover, fixing sewing machines, I see our planet in all its shades and colours; good, bad and indifferent. My customers, who vary in character, shape and form to every imaginable degree, often ask me how I ended up fixing sewing machines. Some, I must add, ask just to be sure that I know what I am doing (as I operate on their precious baby), some to see if I am worth paying and some just to pass the time. One customer, a nun, recently negotiated a discount because, as she forcefully informed me, she had supplied the electricity! I had the feeling that she wanted the receipt written in my blood and a guarantee for life. I swear when the wind is just right, I can still hear her moaning!

I love being an engineer. I am convinced that this world only developed because of engineers. They are the unsung heroes of the human race. No doctor, mechanic, builder or even cake maker would have much luck without the tools of their trade, all made, repaired and often designed by engineers. Take away everything that engineers have built, from castles to computers and we would probably be back in caves throwing bones at each other (which may not be such a bad thing).

Luckily for me, my trade is unusual. In fact there are many more consultant surgeons in Britain than fully-trained and qualified sewing machine engineers. If I wanted to emigrate it is a skill that allows me entry into almost any country, because it is such a rare trade. But the truth is that as we buy more and more brilliantly designed, but cheaply built and disposable products, my old trade, which has taken a lifetime to learn is in decline. That sort of suits me as I seem to be declining at a similar rate.

Actually, I never got into the specialist trade of repairing sewing machines, I just never managed to escape from it. I was born into a manufacturing business that, like so many other companies, was disappearing as Great Britain slowly lost its huge industrialised heart.

I often think back to those precious days of youth. Was my path chosen long before I was aware of it? My start in the sewing trade was bizarre for sure. Like the old entertainer Max Bygraves used to say as he walked onto the stage, faced his audience, and pushed his hands out in front of him like a magician, "I wanna tell you a story".

It all started long before I was born. My very posh English grandmother had run away from home, apparently after a huge family squabble. After many miles and countries, she ended up in Austria, fell in love and married Johannes Faistauer, brother of the famous Austrian artist Anton Faistauer. She then produced my mother, amongst other children, and got caught up in the horrors of the Second World War.

Vienna, where my young mother and grandmother were, was also where my father was studying. In the harsh realities of war my parents met. The city had been ruthlessly liberated by the Russians (I am putting that very mildly) and after 1945 it became an open city, a bit like Berlin, with Russians stomping all over the place, acting as if it was now theirs. Food was scarce and riots broke out. The family made a meagre living running a small fashion house and textiles business, often making garments for the Russian officers and their mistresses. The Russians homed in on my Dad because of his Russian name and started to apply pressure for him to return to the 'Mother Country'. Dad was no idiot and was well aware of what was happening to repatriated Russians. He had escaped once before by the skin of his teeth when he was only three and was smuggled from Moscow to Paris where his French mother was waiting. He had no intention of returning to a communist state. Apparently things came to a head one day when Dad had a furious argument with a Russian colonel. After that he grabbed his family and headed for England, arriving in 1951.

After a few years my parents ended up in Eastbourne. Mum, who I often refer to as 'Mumsie', managed to get a job working in Eastbourne at an up-market dress shop in Carlisle Road, which was half way up the stretch towards the seafront, next to the Lilac Tea Rooms and the long-gone Devonshire Baths. I think that the shop was called 'Peter Norton's' (it was before I was born), and Peter had another shop in Haywards Heath. I was told that Mumsie ran the Eastbourne shop along with her two assistants, Barbara Marples and June Deacon.

The shop was haute-couture: high fashion. As well as carrying out alterations they made all the latest fashions. My Mum, although still only in her early twenties at the time was already a highly skilled Viennese dressmaker. At Norton's she was in charge of the most important jobs, including the cutting-out and final fittings. She would take her children (my two elder brothers), to work with her and the other girls would look after them.

Our town had more than its normal quota of affluent, silk-laced ladies; Viscountess Hampden, the Duchess of Devonshire, and many more. Eastbourne had several high quality fashion shops back in the 1950's, like Madame Jane's in Grove Road, which was opposite Jacksons the Fur shop. There was also Bobby's in Terminus Road, which had five-year apprenticeships for their tailoring and millinery departments, and the best haberdashery in town.

At Madame Jane's their best customer was actually Lady Jane, who kept a suite at the Grand Hotel with her own personal maid. As a trainee dressmaker Doreen Mockford would stare in amazement as Lady Jane spent the equivalent of several months wages on a single dress!

Just along the road a little, almost opposite the Town Hall, was another haute couture shop called Radcliffe's, run by the Stone family who had moved home from Argentina. At the shop a normal day-dress would cost eight guineas, and a felt work-dress, two guineas. Dorothy, who was an apprentice there, would quietly peek between the curtains as customers like Lady Tollemache would flick through Vogue magazines on the table and spend the equivalent of Dorothy's annual wage on a morning's shopping.

Shops like Driscoll's would spend weeks hand-sewing pearls onto evening gowns for royalty. They had made Princess Margaret's bridesmaid dresses as well as other items for the Palace. In one of my books, *Tales From The Coast*, there is a great story about Driscoll's and our Queen Elizabeth. The Driscoll's lived on the corner of Ashburnham Gardens, a few yards from my house at No7. The very prim Mr Driscoll would set off to work each day looking absolutely immaculate, a little like Hercule Poirot. As soon as he was out of sight, I would pedal up to his wall, stand on my pedals and nick the damsons from his tree that he could not reach. Driscoll's shop was a sight to behold, but after Mrs Driscoll died and Mr Driscoll remarried, he seemed to lose interest in his shop.

One day Mumsie had a brainwave while she was walking to work in the rain, wheeling the pushchair. When she got home she knocked up a waterproof cape to cover the pushchair. She called it the Raincape and there was nothing in England like it. So many people used to stop and ask her where she had bought the Raincape that it was not long before she was making them and selling them to customers.

Business boomed, and soon she left her job at Norton's and went with Dad to a rented apartment at the end of Trinity Trees and Terminus Road. This would become his office and her workshop. On the door it simply announced 'Vienna Couturier'.

Their workshops and offices were above what is presently a bakers and tea rooms. It was a large block building, now part of Osborne and Trinity House. They were opposite the imposing art-deco Co-operative building and the Army & Navy surplus stores, which sold a lot of the World War II forces surplus. It is presently T J Hughes. Dad's building was split into several floors, and the second floor was

Permus the hairdressers. They specialised in cheap haircuts, beehive styles and perms, which are now coming back into fashion.

On the top floor was Mrs Storey's, the stocking repair specialists. There, several girls worked on very specialised machinery, repairing the tiny weave of silk and nylon stockings. Each thread was individually repaired at a cost of sixpence per thread or for larger repairs it was one-shilling-and-sixpence in total. New stockings were at least six shillings a pair unless you risked buying them from the market. If you bought stockings from the market, often when you got home and tried them on, you found one damaged or one shorter than the other!

As a sixteen year-old, Jean Cousins worked at Storey's and she mentioned to me that both the Co-Op and Plummers in Terminus Road, also had stocking repair departments. The stocking was pulled over a cup-like device and the new colourmatched thread was shot through the weave with compressed air. It was then invisibly joined to the end of the broken thread. It was amazingly detailed and complicated work, but stockings were expensive and well worth repairing in the 1950's.

The middle floor of the building was where my Dad started his empire building and Mumsie cut and sewed her goods on an old Singer model 201k treadle sewing machine. His previous empires had collapsed twice already, once before with the German occupation in Vienna before the Second World War, and later after the War with the Russians. He was back down at the bottom of the ladder, starting all over again with a new family in a new country. He had already had many jobs in Eastbourne, from bar work in the many hotels, to ice cream sales in the seaside vans to flogging vacuum cleaners. Everyone remembers how hard Mum and Dad worked and Jean commented to me many-a-time, that when she was on a night out in town, she would look up to the windows and see the lights burning where my Mum and Dad were still working.

Times were hard and Dad would cut out the fabric and Mum would sew it up – they were two people with one dream. The orders would come in via the only phone in the vicinity, which was not theirs, but happened to be in the shop on the corner.

Dad, as smooth and charming as ever, had talked the young girl, in the shop on the corner with the only phone, into allowing him to use it for his business. The problem came when it rang for him as he was nowhere near it! The girl on the phone would stall the customers while she sent someone to fetch him from his office. Dad would come hurtling down the stairs, out of his premises and shoot along to the shop to answer the phone, always breathless from his 60 fags a day habit. It was not long before he saved enough money to have his own phone. I doubt if giving up smoking ever crossed his mind. In the 1950's smoking was still considered to have health benefits. As the packets told you, they would calm your nerves, soothe your throat, ease digestion and more importantly, look cool. I cannot remember a time when Dad was not lighting, smoking or just playing with his favourite brand of Embassy Filter-Virginia cigarettes.

During this time Mumsie, was busy making the increasingly popular and now patented Raincape, expanding her product range and her waistline at the same time, as one baby after another seemed to pop out. In 1957, three momentous things happened in my family. The family business, Simplantex was properly formed. Dad was no longer an alien as in October the Secretary of State allowed him to become a naturalised British Citizen, and by far the most important event of the year, I decided to come into the world in a basement flat at 22 Eversfield Road. I ended up as the middle child of six boys.

They soon moved to a small shop along Seaside, which I remember vaguely as a child. It was full of seaside toys: buckets and spades, kites, balls and every conceivable item that a child would drool over. At the back of the shop was a raised area that Mumsie used for sewing. It was always a sore point that I could not play with any of the great stuff in the shop, as although they had enough money to stock the shop, there was none spare to let me have any free toys. However, once a year that changed. On my birthday I was allowed to choose one toy from the shop, as long as it was not the last one in stock.

Dad would buy his materials and plastics with cash and then cut them up, and Mumsie would sew anything that sold. In the summer they did a great trade in fabric windbreaks for windy days on the seashore.

Anything that could be sewed up to make a living was sewn, from leather bike seats to herbal scented cushions stuffed with lavender and horse-hair. Before long the success of the Raincape and other products such as the patented changing mat called the Top 'N' Tail, led, in 1959, to the purchase of a proper factory (the old Bindon's wood yard) in Willowfield Road. Simplantex, later trading under the name of Premiere Baby, grew into one of the largest businesses of its kind in Europe. Dad always boasted that he was 'Mister Plastic' as he had brought the plastics industry to Eastbourne.

Now, while all this was going on, and my parents were incredibly busy, in steps Colin and Vivien Parr. Colin Parr was a burly policeman from the Metropolitan Police in London. The couple became close friends with my parents; I was even page boy at their wedding held at St Mary's Church in Hampden Park. It was my first job and one that I will never forget. I fell into the small pool on the side of the chapel while acting the idiot, something that I was amazingly good at. I was dragged out, sobbing uncontrollably, until that is, Colin waved half-a-crown in front of me. It was the most money I had ever seen. I wiped away my tears, grinned and grabbed the cash! Unfortunately, Dad soon nicked it for his fags and I was far too little to put up any serious opposition.

We were a working family, and I am proud to be working class. Our family tree is littered with wealth and brushes with royalty, and a fair few early relatives were definitely born with silver spoons stuffed firmly in their gobs. However, when I came into this world, all the silver had been pawned and both my parents were working every hour God sent, trying to make a living.

More than once there was little food for supper.

The knock-on effect of these long hours was that the boys grew into a wild bunch of latch-key kids. With little control and no one to clip us round the ear for most of the day, we became a little wayward. With both parents constantly working, we were often only let into the family home when Mum or Dad arrived back from work. This caused a few problems, which came to a head one day when Mumsie had one too many complaints from our neighbours.

The main reason for the complaints was due to our back garden wall. I'll explain. After Eversfield Road, we had a quick stop in a house at 24 Parkway, on the Ratton Manor Estate, with pale blue shutters on the windows. Then we moved to a large dilapidated house at 7 Ashburnham Gardens. It was along the top of a triangle of mainly detached houses joining our road to Prideaux Road and Ashburnham Road, all on the side of a hill. There were two amazingly long dividing brick walls, which many of the back gardens joined. For us these walls were secret paths into other worlds of adventure and danger. They, like the cupboard in Narnia, led through countless gardens and even more delights, like chestnut, pear, plum and greengage trees and the Ablewhite's scrumptious apples; all forbidden fruits but irresistible to wandering children with time on their hands. Stop and imagine for a second scruffy kids in short trousers, one with a catapult stuffed in his back pocket, all crawling along one of the garden walls looking for excitement. We were little terrors.

One day Mumsie snapped and frog-marched us down to the police station in Grove Road. I guess that she used her contacts with Colin from the Met, because she had the lot of us fingerprinted so that if we did anything naughty, there would be all hell to pay. We lined up as a huge policeman took our small fingers and pushed them firmly in the ink then rolled each one onto a pad with separate boxes for each finger. There were no smiles, no laughing, this was deadly serious, possibly the initial procedure before hanging. We were then given a tour of the cells so that we knew exactly where we would be staying if we pinched any more plums from Mrs Haffenden, or greengages from the Flints, or if we snuck into the Sargeant's garden just one more time!

From that point on some of us were better kids. The thought of spending my life in jail was pretty scary. That brought to an end my brief criminal career and I even turned into a saint. One day I found a wallet stuffed with money on my paper round. I happily cycled up to the police station in Grove Road and proudly handed it over the counter, the first good deed of my life. I floated around for weeks with a halo hovering above my unwashed bonce. I received a thank you letter with sixpence in it, which turned my tongue black; not the actual sixpence, but the huge bag of blackjacks that I bought with it. Six old pence and eight blackjacks to-a-penny – 48 delicious blobs of black delight. I was in heaven. That was until I threw up a bucket of black slime that looked like something from a coal mine's slag heap.

Back at the growing business, Dad had a big problem. As a manufacturing firm, cut-make-and trim, he relied on his machinists, piecework-paid girls, just like in the film *Made in Dagenham*. The only difference was that, whereas in the film they were trying to cut the girls' wages, Dad was scared stiff of his women, and seemed to pay them whatever rates they asked unless Mumsie put her foot down. The problem came when the machines broke and the production line ground to a halt. No money, no products, just lots of moaning. Dad would instantly be on the phone, furiously searching for a sewing machine engineer.

Eventually, an old man would arrive, lean over the offending machine with a sigh, then with a deep look of concentration get to work. By the time he left the factory, the sewing machine would be purring away almost as sweetly as the machinist, who had previously looked like she loved to strangle cats for a hobby!

When the machines broke, both the women and my Dad suffered severe fits of high blood pressure, and to this end he came up with a cunning plan.

Dad had noticed that, out of his six rough'n'tough boys, one had some magic in his hands. I could fix almost anything, from a plug to his watch. Why? Who knows, we do come from a long line of watchmakers. Even Mad King George used to have one of my distant grandfathers, Jacques Planché, make him special watches.

So, I was summoned into Dad's big office. In my faded memory of those distant times, I sat in silence in front of his huge leather desk, bought from the retiring Barclays Bank Manager in Terminus Road. I watched him open his letters and read them one by one. Eventually he looked up and asked in his deep voice, "Alex, what do you want to be when you grow up?" That was easy, I had always wanted to be a doctor.

"Doctor, Dad." I squirted out with loads of enthusiasm, revelling at the precious moments I had alone with him. Normally we would be crushed in his car, late for school, or feeding at the table, which was more akin to a trough at Farmer Hoggitts. "I would love to be a brain surgeon." I added.

Dad seemed momentarily taken aback by my quick reply. He leant forward slightly and silently rested his head in his big, tough boxers hands. "Yes," he breathed out with a deep sigh. "I remember now, you once told us that around the breakfast table. I am sorry that we laughed so much, but it was funny for a such young boy to say that he wanted to be a brain surgeon."

Laughed! Yes they laughed. So hard and for so long that I started laughing with them, even though I was supposed to be insulted that the thought of me being a surgeon was so far from possible that it was hilarious.

My father explained that we were, on his father's side, descended from a line of proud White Russians. By his dramatic accounts they were all sword swirling Cossacks, crushing communism with sharp blades and white silk shirts. After the terrible losses of the Second World War, Dad had answered England's call for men, and came here to make a future for him and his family, which I was now part of. He was handed a ten-bob note at customs, patted on the back and told to, "Go forth and prosper." That he did. Now it was my turn to invest my time in the family and the company.

"That is very lucky that you want to be a doctor, Alex, because you are going to practice operating on my sewing machines! I am bringing in some men who will train you, and then you can operate on them to your heart's content. Now pop these letters in the post on your way home." That day I walked home confused, excited and disappointed, all in one.

So, my path was set. Fate had planned out a course for me, one which I would end up following for the rest of my life. However it would not be at the family business. The best palm reader in the land would have had trouble foretelling the collapse of the British manufacturing industry, which would in turn lead to me becoming a master craftsman, running my own specialist business.

Eastbourne was once a hive of industry. Many may remember the huge Jaycee Furniture, or the massive Birds-Eye factories that employed over 2,000 people and ran 24 hours a day.

As well as these big businesses, there was lots of sewing and fabric manufacturing in Eastbourne, from Jarvis Leather Goods, with their government contracts, to Pura Plastics in North Street, which made everything from motorcycle bags to seat covers. There was Lizannes run by Mr Potts, who bought his staff doughnuts most days as they made countless shower caps and curtains for firms like Boots the Chemists. There was Kitestlye, which produced clothes for the London markets, and Coverplas, which made a million tea cosies, tablecloths, and best of all, tiny egg cosies to keep your freshly boiled egg warm. Alston's, in Tideswell Road, run by Tom Alston was a bra, corset and petticoat specialists. There was the world famous Jaeger factory where a dress cost a month's wages. There were numerous fabric and garment manufacturers in the town, and along with the factories, there was an abundance of highly skilled sewing girls who jumped from factory to factory depending on the best wages paid.

The favourite business from my youth was at 62 Susan's Road, it was called Jacqueline's. The shop was painted a shocking pink and sold ladies underwear. Like a good boy I always averted my eyes as

I walked to Chris & Barry's hairdressers next door. Inside, Jackie, who had come down from London in the early 1970's, would get her sewing girls to sew special lace and silk undies. It was quite outrageous back then and she had many letters from the starchy Eastbourne crowd. What they did not know was that, when Jackie took a holiday, the local vicar's wife who worked in the florists along the street, used to look after the shop!

Like so many British companies, most closed because they were unable to compete with subsidised imports, and eventually our business, Simplantex-Premiere Baby, was outcompeted by the massive influx of goods. Only a small part of the parent company survived making wheel chair accessories.

After I left the family firm I started Sussex Sewing Machines, and today I look after countless customers around the South East. No one could have guessed the path that my life would take, even Eva Petulengro, the world renowned Brighton Palmist would have had trouble seeing this future for me. One minute I am advising on the latest Arthur Conan Doyle movie, next helping sort leather machines for the fantastic Bushcraft range of products, headed by Ray Mears the survival expert (some of which Becky at Woodlore makes). Then I am giving expert information to dozens of museums all over the world. My specialist training that had started as a young man in Dad's office all those years ago has stood me in good stead. My father's words had proved right. Although I did not stumble across any Russian relatives riding bareback through the streets of Sussex, I did have a great living at my fingertips.

While I travel around my area visiting the 30,000 or so customers whom I have accumulated over the decades, I often become that fly-on-the-wall, who comes into someone's home and stays for a while, then disappears. The properties I visit range from hotels and farms, to factories and schools. From inner city slums to marvellous manors. I see life in all its shades and often stop to chat – well gossip really. Occasionally I pick up stories that just have to be captured for posterity. They are all real stories and paint a fascinating picture of this strange old world.

My tales grow slowly, much to my publisher's annoyance. I write when I manage to squeeze the time in between all the other messy things that makes up life. Most of my books take around three years to build. I adore writing, like all hobbies there is a great pleasure in it, like that perfect shot on the green in golf, or the first bite on your line at the beginning of the trout season. Everyone who has a hobby knows the pleasure they get from it no matter if it is horse riding or speedway racing, cross stitch or knitting. With writing, I love the way you can play with words to make sentences. Like my poetry, I think about each line and change it again and again until I am happy with it. I build each sentence into a paragraph, and each paragraph slowly grows into a story, and then the stories grow into a book.

In my first few books I used to change some of the peoples' names, but that was a waste of time as they all knew who they were. More than once I have been accosted in the street, "I see you have been writing about me again, Alex!" Or, "Why aren't I in one of your books?" I throw back a few apologies and in my mind's eye, I see the pile of notes in my study that are still waiting to be written, if only I had the time.

So, here we have it. My next collection of anecdotes. If you are looking for continuity and structure blended with a deep plot entwined in linguistic perfection, guess what? You won't find it here, just real people, real life and real tales. These stories come straight from the heart, so please forgive the mistakes, my memory, like my waistline, is not what it used to be.

The title of my eighth book is *Have I Got A Story For You*. You bet I have, loads of them.

BIRTH IN A SEASIDE TOWN

I was dragged into the salt-cut air
Beneath the screaming gull sky, my
Heart pounding in rhythm
With the restless waves.
My tiny lungs suck up the sea-
Flecked spray and my sticky eyes
Prise open. My burning Limbs
Wriggle and my cries of joy
Mingle with the high circling birds.
There is no greater place on earth
To be born – than in a Seaside town.

HAVE I GOT A STORY FOR YOU

Pevensey is one of the most haunted places in the country. Yep it's true. Over the centuries it has been the site of plague and massacre, and of course, the most famous landing in British history in 1066. The Anglo Saxon world came to a crushing end late one autumn in 1066 when a ruthless and cunning duke set his greedy eye upon the fertile Saxon soil. His iron hand and his greedy barons laid waste to a country that had evolved from Roman times. England fell beneath the might and cruelty of the Norman yoke and it all started at Pevensey. Astoundingly, nearly a 1000 years later, the Church, the Crown, and those Norman families still own most of England.

There are more ghosts at Pevensey than you can shake a stick at. Not that I would suggest any such thing as I doubt that it will have much effect. However I'm going to tell you a story about one ghost in particular and the effect that she had on one man. You won't find her in any book, she is a local ghost known only by a few who have lived or spent time in this ancient place.

I heard this story sitting in Chris & Barry's hairdressers in Susan's Road, Eastbourne, though it happened in Pevensey. I had dropped Yana off at Marks & Spencer's and by 6.30 in the morning, I was sitting with a group on the worn vinyl cushions, discussing all the relevant topics of the day and sorting each one out perfectly.

Ian, known as 'Boots' to most of us after an incident in Boots the Chemist that I am sworn never to mention, was sitting by my side. Ian is one of the indispensible and invisible team that keeps our Arndale Centre operating. One minute he is polishing floors with his machine, the next crushing cardboard thrown out by the stores. He is a mine of information and a better forecaster than any London analyst on the state of the economy. He can tell from the amount of waste produced by the countless shops in the Arndale exactly how the town is doing. He is my weather vane on the economic climate and, so far, never wrong.

On my other side was John who looks tough, the sort of bloke that would be useful at a knife fight – short, stocky, and now retired. He used to be a plasterer by trade and pops into Chris & Barry's for a cup of coffee on most days as his routine of early working mornings cannot be broken. John has a lightning-fast intelligence and jumps in when anyone needs pulling down a peg. Sometimes he forgets to turn his hearing aid on and speaks loud enough to disturb the seagulls feeding on the pavement scraps outside. If anyone makes his coffee, it is never right. One day, for a laugh, Chris put four spoonfuls of coffee and six sugars in his cup. John still got up and added more! He is always upbeat and always cutting with his remarks on the economy and football. I have never heard him say a bad word about any person he knows.

Nigel was along the bench a little, a tall powerfully-built man, who works at Hopkins and can fix just about anything. You point at it and he'll sort. Next to him was Graham who has spent much of his 72 years as a ground-worker. If you want a hole dug six feet by six feet, Graham is your man. It won't be an inch longer or shorter. It will be six foot dead.

So there we are, all the usual suspects, sitting waiting for our hair to be cut by Chris, as Barry is on the second shift. The two of them have been cutting my hair since my Dad decided he could upgrade my cut from the Eastbourne Railway Station lavatories!

Surprisingly, when I instantly skip back all those years to when I was a child, I kid you not, there were barbers inside the railway station's loos. There is still a barber's sign outside to this day, but just along from the original loos. Two chairs were bolted to the tiled floor facing the urinals and a small row of chairs by the side-wall for customers. It was a surreal time sitting there as kid, travellers walked in, peed in front of you, while you had your hair cut for a shilling, and walked out again. It could have been worse – the chair could have been facing the long line of toilets behind where more strenuous ablutions were carried out for the cost of one old penny, dropped and turned into the machines bolted to each dark-wood door!

Sitting at Chris & Barry's waiting for my turn in the barber's chair, I was surrounded by the smell of the selected hair potions and aftershaves. On the counter were the tools of the trade, clippers, scissors, razors and a wooden soft-bristle brush for whisking away the neck hair. There were hair clippings on the floor and a small television on in the background. On the table were a few magazines and the day's papers. In the corner is a broom for sweepings, leaning against the red-slatted cupboard door where Chris keeps his guitar and Barry keeps his cricket bat for customers that don't pay! A warm kettle is on the sideboard along with tea and coffee, underneath the sideboard is an assortment of objects, from an old tube-telly to assorted boxes – things that have no value but are just too good to throw away.

John, Nigel, Graham and Ian often pop in for a drink and a chat in the early hours before Eastbourne wakes up, and some go off to work. I was the only one who was actually going to get my hair-cut. Anyway, the bell on the door tinkles and in walks 'Magpie'. Lenny Van Ross earned his nickname of Magpie because, he could spot the sparkle of a deal a mile away. He was a fount of knowledge especially with cars and who was offering anything from the latest deal on hire-purchase to the best servicing. Most of his mates consulted him when they needed to make a deal and he was rarely wrong.

"Here Alex, have I got a story for you," said Chris with a big smile as Lenny took a seat. "This will make you laugh." Chris always astonishes me with his amazing amount of local knowledge; it makes little difference what topic, from music to boxing, he will add something that I never knew. Someone once said that if you wanted to know something ask the barber, and at Chris & Barry's it was true. In fact we laughed that my next book should be called *Tales From The Barbers Shop*. I love that idea and just might use it.

Anyway, Chris then started to relate the tale of the Pevensey gargoyle which involved Lenny. Every few words Lenny, John and Ian (who had heard the story a dozen times) all jumped in, correcting Chris, so that the story was spot-on.

I was sitting in the sunken comfortable barber's chair looking at Lenny and the lads in the reflection of the large mirror which had business cards and postcards stuck around its edges. One read, 'Smile first thing in the morning and get it over with'. After a few nudges and interruptions Lenny took over telling the story from Chris.

"Well it was like this. I was working for old Ray King, you know the night club impresario. His jewel in the crown was King's Nightclub down the Crumbles, long gone now but in its heyday, well, it was impressive. I worked for Ray King as general help behind the bar, serving, or whatever needed doing."

As Lenny talked, images of King's Nightclub sprang to my mind and fond memories rushed back from my youth. In the 1970's the place really was impressive, seating for hundreds with at least a 2,000 capacity. Every table had a telephone in the middle in case you had to make a last minute deal while watching a show! They were the old telephones with the large rotating dial in the middle and a handset big enough to knockout a burglar with. All the cables were neatly hidden away. They had a bar that could dish out vintage champagne alongside cheap beer. Occasionally Ray, always dressed to impress, would move smoothly through the room like some East End mob boss. He had an instant magnetic presence. King's was the Mecca of the area and attracted every big act of the day. I remember Prince Charles coming down to see the Three Degrees, (who he openly adored). I saw a few big names there myself, the funniest by far was the brilliant comedian Tommy Cooper, but not for the reasons you may imagine.

If my memory serves me right, which is no guarantee as I was mentally scarred for sometime after the event, my brothers and I had gone to see Tommy for a laugh. Before he came on Ray King had booked a stripper to warm up the crowd and fill in some time as Tommy liked to take to the stage late. Well, what were we going to do? We took it all in good fun and sat there while this fat old woman (at least 35 anyway), came on stage and started stripping to the standard heavy Ba-ba-bum-de-baba-bum from the live band. We drank our drinks and cheered along with everyone else, waiting for her finale. All was going well until someone started heckling the big girl as she dragged out her performance. Now, she was a tough old bird and after just one jibe too many, she held up her hand to the musicians and the music stopped. There was loud cheer and applause. Half naked with the most enormous boobs I had ever seen, she got down from the stage and moved towards the heckler, who was behind me. Wow was he going to get it!

She could not see well with the thick haze of smoke, and summoned the man working the spotlights to follow her with the lights.

My brother Sam was loving every second, cheering along with the rest of the room. As she got closer, sashaying round the tables, the spotlights moved ahead of her. Sam, always the joker, instantly saw an opportunity. As the dragon lady got nearer Sam suddenly stood up and shouted, pointing directly at me, "Over here, here he is."

Well, before I knew it, the spotlight stopped on me and I was looking up at a sight that has shaken me to this day. My innocent teenage years were about to be blighted forever. "Like to heckle do you love?" Were the last words I heard... and saw, as she reached round behind my head and pulled me, with the force of a Welsh Miner, into her vast bosom. It was like being held underwater. I could not breathe, I was aware of muffled cheering through the amazing sound-proofing of my new earmuffs.

Shock turned into blind panic as I realised I could not suck any air in at all! My arms flayed around spilling my beer as the crowd laughed themselves silly. After what seemed to be an hour, and just as I was about to drop dead, she released me. I fell gasping back into my seat. "Be more careful now won't you love!" She said as she sashayed back to the stage with rapturous applause, whooping and cheering.

The band then struck up and she continued—with not a single heckle!

I was the colour of a fresh boiled lobster. I panted for breath and looked up to see my brothers wetting themselves with laughter. My face was covered in make-up sparkles, and through sharp intakes of breath I shouted a few perfectly descriptive words at Sam. The real heckler behind me came and thanked me for saving his life and happily patted me on the back hard enough to re-start my heart.

I cannot remember much of what Tommy Cooper was like, as the rest of the evening was spent in sporadic outbursts of laughter each time I looked at my brothers.

The best night of the week at King's was Thursday night, which was single's night. Unfortunately as divorce in the 1970's rocketed for the first time in history, single's night was taken over by older women looking for lost love. It was not long before Singles' Night at King's became nicknamed Grab-a-Granny Night.

King's Nightclub was just one of the establishments that Ray King owned. He would drive to his pubs and clubs in his gold Rolls Royce from his large mansion on the Ersham Road called Glyndley Manor. I am sure that the number plate was something like Ray 1.

Ray had a pub in Pevensey called the Royal Oak and Castle, right in the shadows of the famous Pevensey Castle and that is where Lenny's story continued. Lenny worked for Ray, and was told to look after the evening bar; much to Lenny's better instincts, he agreed. Lenny was well aware of the ghost stories of Pevensey and would make sure that he was not left alone at night in the pub.

As each evening went on he would make sure that someone stayed with him while he went round and locked all the doors before he got into his prized Austin Allegro and shot off. He hated checking the upstairs rooms, as one room had a rocking chair that insisted on rocking by itself and a loo chain that swung with no reason.

Some say it was the old lady who went mad after witnessing the grisly murder of the lovers in the Mint House opposite.

The Mint House is an ancient crooked building built on the sight of the Pevensey Mint and was once the home to one of King Henry VIII's physicians. The story goes that one night in 1586 a London merchant returned home unexpectedly. Thomas Dight came back to find his mistress in bed with a lover. In a demonic rage he murdered them both in horrendous ways. Locals say that the Mint House was haunted from that day on.

The old lady in the pub across the road, who witnessed it all from her window, went mad. She now haunts the top rooms of the Royal Oak and Castle. However they are not the only ghosts at Pevensey. There is the Grey Lady, supposedly Queen Joanna of Navarre, who haunts the castle paths, and the Little Drummer Boy who plays out his drums along the outer walls of the castle. Then there is the sad little girl who stands sobbing over the Plague Pits behind the church, and many more.

All these tales of ghosts left Lenny decidedly jittery, so when he locked up the old pub he would try and get at least one of his mates to stay, even if it meant buying them drinks, until he could make his escape.

One day, his mates came up with a devious plan to get Lenny, good-and-proper. They knew Lenny always parked his car under the yellow fluorescent street light by the pub, so that he could get to his car with some light. In the driver's foot-well they placed a large plastic owl with huge reflective eyes, like one of those used to scare birds away from allotments. They angled it perfectly so that when Lenny opened his car door the light from the streetlamp above would reflect in the owl's eyes and give Lenny the fright of his life. Then, all they had to do was to make sure that they were hidden away to see what happened.

As the evening in the pub wore on, Lenny's mates, one-by-one, made their excuses to leave. By locking up time even with all Lenny's pleading and free drinks, no one was left. Lenny gripped himself and went upstairs to make sure everything was closed. He rushed by the room with the rocking chair and made it back down to the front door.

Lenny locked up the pub and walked quickly over the gravel car park towards his car under the yellow street lamp. The castle walls looked black and menacing in the dark. As he walked, unknown to him, his mates were hiding behind the bushes and throwing stones behind him to scare him. Lenny kept looking back and rushed to his car cursing all the way. He opened the door and there staring up at him was this gargoyle with yellow devil eyes. Lenny screamed and fell back from the car.

His mates muffled their laughter as he then proceeded to grab handfuls of gravel and pelt the car with stones, trying to shift the gargoyle and get it out of his car and back to one of the pillars that it would normally inhabit. Lenny became more and more furious, shouting at the beast but also because of the scratches he was making on his freshly washed car. When Lenny had calmed down from his initial shock he edged closer to the car to get a better look. It was then that he heard the laughter coming from behind the hedge. Instantly he knew what had happened.

As Lenny's friends scarpered into the night, Lenny threw the owl out of his car and raced off in disgust shouting a few words of explicit instructions out of the car window as to where his friends could go. The whole tale, and how Lenny told it, was brilliant.

So there I am in the barber's chair laughing my head off. Chris had to stop cutting and was leaning against the counter for support. "Aah, I've heard that so many times but it still cracks me up," wheezed Chris through his teeth. The lads on the bench were also having a good laugh, and even Lenny, who had told us most of the tale, was having a chuckle.

Chris was right, he did have a story for me, and what a cracker it was. I left the shop still smiling imagining the shock of that yellow-eyed devil in Lenny's car parked at one of the scariest places I know.

As I walked down Susans Road I turned back to look at Chris & Barry's, the best barbers in town. In the sepia sunlight I see my Dad with his hand on my young shoulder walking me up the street to get my first short-back-and-sides at a proper hairdressers. No more station loo's for me! I am about nine, in my summer shorts, with a round happy face and so excited to go to a man's barbers. The bell hanging over the door announces our arrival and we disappear.

BABYLON WOODS

The dark woods are broken by small clearings of light where old fallen trees have left holes in the dense canopy. Grey beams of light pour through, falling upon the ground where a few ferns and bushes eke out a meagre living on the soil and sunlight.

The tree-cover is dense on the steep hillside in Babylon Woods and hardly a plant grows. It is a primeval sight that I have known since my father walked his six sons here to sap their boundless energy.

In those days of youth, we searched for old mosaics that Dad promised us were from a roman villa that was once cut into the hillside, where a roman general surveyed his great lands of Anderida. Now I wonder if we were just gullible kids or if Dad knew something? Without the forest it would have a spectacular viewpoint. If I had only kept a few of the small square blue mosaics we had found I would now be able to find out.

The path snakes through the trees curving away from me into the darkness of the cloudy morning. On my left the woods rise sharply as far as the eye can see, and to my right they fall away to houses hidden by the trees below.

Hardly a breath of wind moves the leaves in the canopy of yew, ash and beech. Babylon Woods must have one of the finest gatherings of yew trees in the area, maybe in the country. They are ancient, tall and spindly, reaching up in twisting gasps desperate for light and life. Some have fallen but in a last moment, refuse to die and curve back upward, creating unusual art forms of shape, strength and power.

Rolly, my little black devil-dog, is hunting, cutting the path left to right then back. She stops for an instant and glances toward me, then she is away following every scent, every track. Although she is now old she is as fit as a puppy, and moves with fluidity as she chases shadows and movement. Her blackness blends with the darkening sky above and I lose sight of her.

Above me there is just enough wind to make the long-legged sinewy ash slowly sway. In the canopy they move like sleepy dancers in a nightclub touching, caressing. As the leaves brush together they make delicate noises like lovers' whispers. Their slow sensuous dance is hypnotic and I watch the enchanting performance as if I had been a privileged guest.

Invisible woodland birds sing cheerfully as if excited, as if something is about to happen. No one is here, I am alone with nature. A hundred thousand people are in the town below and I am here by myself.

I stop as movement catches my eye. I see a leaf move, then another. Rain is coming. The heavy summer dawn has brought much needed water. I look on as more leaves are hit by large raindrops. The leaves dip and then spring back like piano keys. Babylon Woods starts to play a tune. It is the very essence of forest life. Countless raindrops fall and the flat leaves of the ash and beech become frantic with the increase in rain. All around Babylon is alive with sound.

I shelter under a bent yew and watch the orchestra as it reaches its crescendo. The noise is amazing, deafening. Then a flash of lightning pulses through the trees reflecting on the wet barks as it streaks by. A peel of thunder follows, deep and lingering in the curve of the hill that acts like an amphitheatre. Rolly stares up at me, her big brown eyes look on with fascination. She is in her element. She shakes the water off and rushes away.

As the rain eases it is replaced by ponderous droplets that have accumulated on leaf and branch. There is no hiding place now and I get soaked. My cashmere jumper is wet, warm and wet.

I decide to move as there is no point in trying to shelter now. As the last rain runs away to hide in the earth the quietness returns. The leaves take their final bows and the woodland orchestra slowly fades. There is no birdsong now in this primordial scene. A heavy silence mingles with the scent of warm forest floor and the wild marjoram drifting in from the surrounding downland. It is delicious and totally intoxicating.

Babylon is breathing. The very earth is breathing, deep and slow. A mist rises from the warm soil as she sighs. Babylon Woods once more trails moisture back to Heaven. The mists collect just above the tree line and forms light clouds that linger for a moment creating a miniature rain forest in the summer heat.

A pigeon announces its presence to the world. Her voice echoes hauntingly around the hillside and is lost deep into the distance, a woodpecker drills its mating message high above me. Babylon Woods has worked her magic once more.

I get back to the Land Rover, Rolly jumps in leaving muddy paws all over the grey leather. The mysterious forest calls to me as I turn to leave. I look back into the darkness and I know I will return. I have been drawn here year after year, decade after decade. Babylon Woods entrances me.

She is in my blood, in my very soul.

THE CAVENDISH POLTERGEIST

"What are you doing girl? Give me the covers back."

"You did it, you twit! Now go back to sleep...we have work tomorrow."

That's how it all started way back in 1966. Maureen and Doug Watson had moved down from London, along with Maureen's parents to find work in the bustling holiday resort of Eastbourne. They moved from Brixton to an upstairs flat at 70a Cavendish Place, almost on the corner of Tideswell Road. They had agreed to rent the property from Mr Dallaway, the Pevensey butcher, (that doesn't sound quite right, but you know what I mean). He made a bit of extra income on top of the meat trade by renting out a few properties. He never mentioned anything unusual about the flat to his new tenants! But there was something very strange about the attic room at 70a.

The property is a substantial three-story town house probably built in the late Victorian period. It is a terraced building running along Cavendish Place. It was built to a typical high Victorian standard with high rooms and extra ornamental decoration inside and out, and a small balcony running along the entire front of the premises. It was a perfect town house, though today looking decidedly squalid. Right at the top of the property was an extra floor and attic room with a small dormer window looking out over All Souls' Church and the old centre of Eastbourne, much of which was swallowed up by the Arndale Centre. It was in this dusty unused attic room that the Cavendish Poltergeist came to visit.

The Watson's first night at the flat was to be the beginning of a weird and fascinating journey into the unexplained. They had put their bed up into the small attic room, which had the best views over the town. They settled in and fell asleep.

Without warning, the window opened and bedclothes flew off the young couple.

"What in God's name, Doug?"

"Wasn't me," he replied, sleepily getting up and heaving the heavy woollen blanket back onto the bed.

"If it was not you, then who was it?"

"Love, I didn't do nothing, you must have left the window open. I was fast asleep. We'll talk about it in the morning."

They put the incident down to each other and fell back to sleep. The next night exactly the same thing happened. The window flew open and the bed clothes lifted off the bed!

"Maureen love, we're moving!" Said Doug standing bolt upright in his pyjamas.

"Where to? It is perfect here so close to town, and I am only ten minutes walk from my work at the Cavendish Hotel. Let's just move downstairs to one of the rooms below and see what happens tomorrow."

And that is exactly what they did. In the middle of the night they clambered down the stairs with their bed to the room below, set it back up, and slept like babies. Not a single disturbance. However, things were about to change.

One day, Maureen made a nice display basket of flowers and put it on the sideboard. When Doug arrived home from work she asked him what he thought of her display.

"What flowers dear? There ain't no flowers 'ere."

"On the side...can't you see them, have you gone blind?"

"Well knock me down with a feather. Where are they?" said a baffled Maureen as she came out from the kitchen where she had been preparing Doug's supper. "I swear they were here not thirty minutes ago."

They searched the flat then both looked at each other and glanced towards the attic bedroom. Slowly they made their way up the stairs and Doug opened the door. On the dresser was her basket of flowers. "Strangest thing I ever did see," whispered Maureen, not wanting to get too close to them. "I think I'll leave the flowers here for now," she said, quickly leaving the room. "I'll just check the potatoes love."

Doug had grown up in London with his grandfather who had emigrated from Sweden years before. His grandfather had taught him never to be afraid of the unknown. "There are more things in heaven and earth than we shall ever understand," he used to tell his grandson when looking after him. "Never be afraid of the unknown – till you know there is something to be afraid of. Dougie my boy," his grandfather would say. "If the hair on your neck stands up and every nerve in your body tingles, then pay attention, something amazing is about to happen!"

Things quietened down for a week until Maureen's lizard brooch disappeared and Doug found it on the dresser, next to the dead flowers, in the attic. This time he locked the door and put the key in his pocket. The next day Maureen told him the brooch had gone again. They went once more to the attic and Doug quietly unlocked the door. Sure enough on the dresser was her brooch. Maureen took it back, looked around the empty room and left. The next day it disappeared again and search as they did it was never found!

"We need to do some investigating love," said Doug over dinner. "On my day off I'll pop up to the Town Hall in Grove Road and the Records Office to see if they have any details of the previous owners. I bet the local paper archives in Commercial Road will have some old details, and the library. We'll get to the bottom of this, you mark my words love, we'll suss it out."

"Good idea, I'll ask around here to see if anyone has heard of anything similar like this happening before," replied Maureen.

The next night they sat down and compared notes. "Well I found something out that you are not going to like," said Doug with a worried frown.

"And me," Maureen piped in. "You first Doug, what did they say at the Town Hall?"

"Well, we looked through records right back to the last century and blow me down, back in 1893 a girl died in suspicious circumstances up in the attic. They say it was probably carbon monoxide poisoning from newly-fitted gas lighting, which may have been faulty. They say that if the window was open she would have been fine. But, and it is a big 'But' they are not positive and they did not even have her name, which was weird. All they could tell me was that the girl may have been a new immigrant, which is possibly why she was not registered properly? It was all rather sketchy, but the lady who helped me said she would write to us if anything else turned up.

When she asked why I was so interested in the history of the flat I was a bit stuck. I couldn't spark-up that we've been 'avina-haunting could I? They would take me straight off to the funny-farm. Anyway, that's my day. How did you do love?"

"Oh well, Payne's the Convenience Store downstairs on the corner, told me something strange. Freddy Payne was so interested he actually stopped his trumpet playing in the back store and came out to talk to me. Apparently the top of his building and ours were once all part of the same large attic. He reckons that there was a tale of a girl who lived up there. She would stand on a chair and stare out of the window for hours. He is so worried about going up to the attic, in his part of the building, that he won't put anything there. He told me that it often ended up broken or useless anyway. He said that the owners of the shop below us, the engravers, Keeley & Alcock, never go upstairs. Even when the roof leaked they paid someone else to take a look!

There's more, and this is a cracker, apparently a priest was called to bless the building but he ran off when his torch went out. They found him shaking like a leaf in The Hartington across the road. The barman said he downed three double whiskeys in the time it took to pour them, and refused to return to the building!"

"What are we going to do about it love?" asked Doug, munching on his supper, amused by the whole situation.

"How about us getting a priest as well?" replied Maureen, with a deep look of concern. "Perhaps they could excommunicate the ghost!"

"You mean exorcise it love. I don't think the ghost would be too bothered about being excommunicated by the Pope!"

"I know! Let's invite Mum and Dad up for supper and talk to them about it. They may know what to do."

"Great idea love. I'll do it after I've finished me pie and mash. Anymore of that lovely gravy?"

A few days later Maureen's Mum and Dad came to supper. They did not have to come far, Emily and Ernie had moved down from London with Doug and Maureen and were living in the flat below, completely unaware of any strange goings-on in the attic.

"Load of cod's wallop, Son. Ain't no such thing as ghosts! Believe me, if there were ghosts I would have seen a good share of them during the War. Nonsense, that's what it is. Nice gravy Maureen love, did you use beef stock?"

Maureen did not reply, but froze with a look of horror as the bottle on the sideboard floated across the room and dropped straight into her Dad's glass with a loud plop!

"What in heaven's sake are you doing Ernie?" shouted her Mum. "Why did you do that?"

"It wasn't me dear, the bottle was on the shelf and I was nowhere near it."

"Who's playing the fool here? Is this some sort of joke you kids are playing on us?"

No sooner had the words ushered from Emily's lips than her chair, with her in it, started sliding across the floor. She screamed and leapt out of it.

"Blimey O'Riley love, let's get out of 'ere," Ernie squealed in a high-pitched voice. "I don't think we will be staying for pudding. We have to get a move on. Things to do, places to see, dogs to feed. Come on Em', were leavin'."

"You ain't got any dogs," laughed Doug, still eating his supper. "Anyway you only live downstairs. Believe me now, don't you?"

As Maureen's parents left, Ernie leaned over to Doug and whispered to him. "Get a priest in here quick son. Things ain't right." He then ducked and high-tailed it out of there just as a coat hanger flew across the room at him.

"Blimey, she didn't like your parents much did she?" Said Doug, laughing his head off.

"No, how strange," replied Maureen with a bemused look. "She hasn't been so bad to us but I suppose we haven't been rude to her. Where I work there is a man who is a clairvoyant I might speak to him next week. I'll see if he'll listen to me and maybe bring him home."

"Good idea love. Now let's tuck into all this lovely grub. No need for it to go to waste eh, great spuds!"

They both sat back down to finish their supper and kept laughing as they thought about what had just happened. They could here the muffled sounds of Maureen's parents down stairs talking ten-to-the-dozen.

Neither Maureen nor Doug ever felt threatened by their poltergeist. In fact to them she seemed quite friendly, if not a little mischievous. The worst that had happened to them was that the bath taps turned themselves off now and again when they were trying to run a bath, and the attic window would regularly open, even when there was no wind.

Maureen talked to her colleague at the Cavendish Hotel, but before the clairvoyant could call round Maureen's aunt and uncle came to stay. They were told nothing about the strange goings on. Their week's visit to sunny Eastbourne got off to a good start. However without any explanation George and Carrie packed up and left after just two days and never asked to stay again. When they did visit Eastbourne, they made any excuse not to call, staying at a hotel and meeting the family out.

A while later, the clairvoyant, who was well-known in Eastbourne during the 1960's, paid Doug and Maureen a visit. He went up into the attic room and tried to talk to the presence. The room went icy cold and he turned deathly grey. "I need to go," he muttered, as he pushed past the couple and headed for the flat door. Later at work, he explained to Maureen that there were phenomena that even the best scientists could not explain. Einstein said that the distinction between the past, present and future is just an illusion. The clairvoyant had clearly felt the presence of a young girl who would not tolerate old people near her. He said she seemed happy with young people but he got the distinct feeling that she disliked older people and would remove them any way she could. He suggested that, as the girl liked Doug and Maureen, they should try and communicate with her. The first thing to do was to try and find out her name. One of the tricks that he used to do was sprinkle talcum powder over all the surfaces of the haunted room and see if there was any movement, especially the floor as it stopped any tomfoolery. With the floor 'talced' it would be impossible for anyone to move around the room without leaving a trace.

"I've bought some talcum powder Doug," shouted Maureen as she opened the door to the flat. "The clairvoyant tells me it might give us a clue," she added as she unloaded her shopping. "I am not sure what will happen but would you be a dear and sprinkle it all over the attic room. I don't fancy doing it myself as I know I'll have to clear it all up again, and that useless vacuum cleaner you bought seems to throw out as much dust from its fabric collection bag as it sucks up."

Doug did what he was told and carefully applied talc all over the attic room, then locked the door. "Key is in the drawer love. We'll have a look in the morning and see if there has been any movement. There will probably be rats footprints across the floor and much moooore," he said in a creepy voice, as he chased Maureen around the room with a ghoulish chuckle. The next morning they walked up the stairs and slowly opened the door. In the bright sunlight all was calm. A slight haze was in the room as the talc caught in the morning light. There was the fresh smell of fruit in the air and a happy feeling. Doug examined the floor closely for any signs. "No footprints, not even mice!" He announced with the air of a detective hard at work.

They looked all around the room for any other sign of movement; neither of them was expecting to see anything.

"Doug, Doug look here," whispered an astonished Maureen summoning Doug over with frantic waves of her hand. She was pointing at the dressing table where a single word was clearly spelt out in the talcum powder. It was a girl's name but with an unusual spelling, Suzzanne.

"Well I never, at least we now know her name. I wonder if we can do something else. Ask your friend at the hotel if there is a way we can communicate with her. Time for breakfast love," said Doug patting his stomach. "Should I lock the door anymore?"

"Why bother Doug, she seems to stay up here unless old people come. Leave it for now. I don't like the idea of her trapped."

"Silly girl, I don't think our non-paying guest, Suzzanne, is trapped. If anything she seems to be able to move straight through from one side of the building to the other. Now where's my bacon and eggs? All this excitement has given me the appetite of a horse. Blimey our own poltergeist, won't your dad be furious!"

The very same day Maureen came back from work with a little glass bell. "Look what I have bought. Apparently spirits love bells! It is the perfect way to communicate with Suzzanne. Shall we give it a try? I am so excited, let's do it before its gets dark. I don't fancy upsetting her and having our blankets pulled off again."

Once more the couple went to the attic and carefully strung a thin piece of string from one of the wooden painted roof beams that came through in the corner of the room. On the end of it they tied the small crystal bell. They stood back.

"Don't move, hold your breath love." Doug gently spoke into the air, "Suzzanne, are you here?" 'Ting' went the bell. They both stared at each other with saucepan eyes. "Are you happy here?" 'Ting' went the bell again. Doug and Maureen moved close to each other for support. "Do you like us?" 'Ting' went the bell once more. They both smiled and Doug gave Maureen a squeeze, breathing a little sigh of relief. "Do you like old people?" Ting, ting, ting, ting the bell rattled in annoyance.

"That's pretty clear," whispered Doug. "And it explains why we only have problems when older people are up here. Are you happy for us to stay here?" He asked to the air with raised eyebrows. The bell rang out once with an approving 'Ting.'

From that incident the months rolled by. Each time there was a little incident they knew Suzzanne had been up to her old tricks. There was no rhyme or reason as to when something happened, no certain times or dates. However, both Doug and Maureen would always smell fresh fruit in the air before something bizarre occurred. It was always the sign of her presence. Eventually the attic key disappeared from the drawer, but they had stopped locking the room anyway.

They never found out if Suzzanne was the little girl who died from the faulty gas lights, but the years shot by, and, when Maureen's parents moved to a council house they took on the lower floors and opened a Bed & Breakfast. Maureen's experience gained at her hotel job had paid off and together they ran the B&B for many years. They always made a point of stressing to their guests not to go up to the attic room as it was empty and there was absolutely nothing in there.

Only once did a guest leave unexpectedly after being nosey and visiting the attic. The old woman hurriedly packed her bags, explained between gasps, that she had felt a draft and went upstairs to see if a window was accidentally left open because it was so cold. Doug and Maureen looked on smiling to each other as she started to explain further but then she got into such a fluster that she stopped herself. She had realised that what she was saying would normally sound ridiculous. She paid in full and almost ran out of the building, dropping bits of her luggage and umbrella as she went. Doug and Maureen laughed so much Doug had a sore throat for a week. They guessed that when she went up to the attic she had probably bumped into their 'spooky' guest.

I called on Maureen and Doug after the couple had retired to Northbourne Road and while I was fixing their sewing machine they told me of the strange goings on at 70a Cavendish Place.

This was a real story told to me first-hand by the people who had actually witnessed the events. They are as real as you and me. When Doug did more research he found that the corner shop used to sell fruit and was owned by a Turkish immigrant and his family who suddenly moved away. However, he never found if it was his daughter who haunted the attic but it might have explained the smell of fruit when their spectral guest visited.

The building is still there, though no business seems to survive there for long. You have to wonder if Suzzanne has scared more than one owner away.

When Doug and Maureen moved, after they had cleared everything out, they went up to the attic room to say goodbye to Suzzanne. The room was warm and bright with a powerful smell of fresh fruit. Over many years they had become very familiar with her and they explained why they were moving. "We used to tell the girl everything you see," said Doug to me over the table. "Like bee keepers tell their bees. That way everyone was happy. We told her that she could keep the lizard broach and the attic key. Although the bell did not ring we both felt a calm happy feeling come over us as if she was also saying farewell."

One final point that they mentioned to me. When they left, they forgot to take down the little bell that was hanging from the beam. I often buy my morning paper from the little newsagents in Cavendish Place and glance up at the attic of 70a. I can't help wondering if, every now and again, the window flies open and there is a 'ting, ting, ting' as the little girl that smells of fruit comes-a-calling.

BROKEN DREAM

Our step quickened to match the drummers' beat,
Excited breath clung in the autumn air.
We marched as one, along the cobbled street.
Boys peeked around door frames, eyes all a-stare.

Up the dew kissed fields toward Tyler's Reach,
Across the meadow sweet to the ridge ahead.
With bible high, loud words the preacher preached.
He saves the souls of us in royal red.

The first rays of sun brush our nervous faces,
Thoughts of loved ones swell our racing hearts.
On the hill they lay, just a few more paces,
The drums quicken and bagpipes play their part.

Men faced men, proud strong men, men of war,
Countless would die, that was the price to pay.
A thousand screams, we rushed into the fore,
A wave of death fell on a wall of grey.

To die for honour we would not think twice,
For God and glory to prove our worth.
Though tear stained petticoats tell the real price,
And childless widows toil the lonely earth.

Bang, bang, went the guns, the shouts and the roar,
I rose all a sweat, my dream now broken.
Wrenched back from a time that made the heart soar,
A hell from which I had thankfully woken.

THE DAY THAT SUSSEX DIED

"Alex, we have a common interest you know!" Said Hillary, as she handed me a steaming cup of fresh coffee.

"Now you have me guessing," I smiled. I couldn't imagine going to quilting classes or dance lessons. Perhaps she has mistaken me for one of her cooking buddies? Just the thought made me shudder.

"It is your wife's family and my husband's, they both had relatives in the Lowther Lambs. In the book *The Day That Sussex Died* both names are nearly together. I'm a Richards and your relative was a Reed." Hillary then disappeared and came back with a large book, shuffling through the pages as she entered the room with her head down staring into it. "Here it is – a list of the men killed at Boars Head on the Western Front in 1916."

All of a sudden an old family secret came rushing back from the distant past.

I had been idly perusing photographs at Yana's family home. I was a spotty teenager, typically bored but making a bit of an effort to put on a nice face in between muffled yawns. It was a lazy Sunday afternoon and the mother-in-law's famous Sunday roast had slipped down a treat. Staying awake in the warm living room was proving a mighty task and, as the new boyfriend, I was trying to impress the family with my interest. "Really, another picture of you on a picnic, how sweet. And Yana really made that dress, no!" A photo slipped from the back cover of the album. I picked it up and before me was a young man dressed in military uniform. I had never seen it before, his deep hollow eyes stared out at me, calling to me from the past. "Who is this?" The photo was immediately passed to the oldest member of the clan, the oracle on all old photos.

Granny Iris sat up and straightened her wig, pulled her glasses up that had slipped to the tip of her nose and examined the photo. "That is Nelson," she said adjusting her false teeth so that she didn't hiss at the end of each word. "He enlisted underage and went to his death in Flanders. You know the poem, 'In Flanders Fields the poppies blow, between the crosses, row on row'. Well that is where Nelson is. I was always angry at him for going away. They never found his body." And with that, the photo was handed back and Iris slid back down into her chair. The oracle had spoken.

I took the photo home and that started a search into a fascinating family ghost. Over many years I talked to old family members who had tiny snippets of memories of him and researched details of his short existence. What came to life was one of the most amazing tragedies of the First World War, which the War Office of the day hardly bothered to even mention. It was a battle where heroes were made and many young men became no more than forgotten whispers in the wind. It was later to be called The Day That Sussex Died.

In September 1914, Colonel Claude Lowther, owner of Herstmonceux Castle, formed the Lowther Lambs regiment and set up recruitment offices all over Sussex. The Sussex lamb was their motto and mascot. Peter, the actual lamb, survived the war and was buried at Herstmonceux Castle in 1928.

Nelson John Reed was living with his parents at 10 Eshton Road, Eastbourne, when he heard Lord Kitchener's call to arms. His father, Jack, had been in the Light Horse Infantry and was proud that his son wanted to enlist. Nelson went with his mates to enlist at the Terminus Road Recruitment Office, however the queue was so long it stretched all the way along Grove Road almost to the Town Hall. Nelson eventually enlisted in Bexhill, he, like so many young men, was eager to fight for his country. The volunteers, who were almost entirely from Sussex, formed the 11th, 12th and 13th Royal Sussex battalions of the 2nd Southdown Regiment. Over 600 men from the 12th battalion were friends and relatives from Eastbourne, especially the Seaside end of town, and the 12th soon became known as the 'Pal's Battalion'.

The men were initially based at Cooden Mount Camp, Bexhill, where they received some of their training from recently-promoted Sergeant Major Nelson Carter. The mood was optimistic and cheerful and even a song was penned to Colonel Lowther, called *Lowther's Own*.

Oh, the Sussex Boys are stirring in the woodland and the down,
we are moving in the hamlet and stirring in the town.
For the call is King and Country, since foe has asked for war,
and when danger calls for duty, we are always to the fore.

Nelson Reed and Nelson Carter were local lads. Nelson Carter's parents, before moving to Hailsham, lived in Latimer Road and Nelson Reed's in adjoining Eshton Road. Before the war Nelson Carter had walked from his home at 33 Greys Road in Old Town, past St Mary's church, where he had married Cathy and had his daughter Jessie baptised, to work as a doorman at Old Town Cinema, along the High Street, near the Old Star Brewery. It was renamed the Regent Picture House in 1921 and then the Plaza. All the local lads looked up to and respected the older, highly experienced, Nelson Carter. He was a tall powerful man and had tattoos from his previous trips with the army abroad, one saucy one and one of Buffalo Bill who he saw when his troop visited Eastbourne. Some of the lads were almost children, like John Searle in the battalion, who was only fourteen when he enrolled!

The battalions travelled to France in March 1916 where they received further training and eventually dug into their trenches at Richebourg. Here bombs fell, rats nibbled at their rations and gas came across from the enemy. Morning rum rations were issued and letters were written home disguising how bad conditions were. Snipers strained for any poor soul that peered over the trenches. Drinking water was carried across in petrol cans stinking of fuel. No baths or clean clothes. The reality of war was biting.

Many people know July 1st 1916 as the start of the bloodiest event in military history but it was in fact Saturday 30th June that the battle of the Somme really started with a diversionary attack to the north at Boars Head (so called because of it shape), Richebourg, a few miles from Lille and the part of Northern France, Belgium and Netherlands traditionally known as Flanders.

THE BATTLE

The fateful day on June 30th started with an hour of sustained bombardment of the enemy. As Sussex went 'over the top' to the sound of whistles blowing, it was to be Nelson's first and last military action. Still dark, early in the midsummer morning the flares went up to signal the attack. Thousands of men shouting, "For King and country," went up the ladders towards the enemy and into the barbed wire mined, nightmare of No Man's Land. Nelson and the 12th Battalion had orders to attack the front line at Ferme du Bois.

Sadly, security was virtually non-existent in those days and the Germans knew exactly what they intended to do. In fact just before the attack they were shouting, "Why are you waiting Tommy, we have some surprises for you. Come and get them Tommy!"

As they came out of the trenches, just as Lieutenant Colonel Grisewood, commander of the 11th had predicted, they were mown down by rapid gun fire. Before the action started, when he became aware of his orders, he had refused to send his men to slaughter and been summarily dismissed.

The Germans had positioned machine guns to fire diagonally across each other making a curtain of death that few would survive. After the failed "surprise attack," the men tried to get back to the relative safety of their own trenches but were targeted by German artillery and blown to pieces.

Most of the officers in Nelson's battalion were killed in the first attack, leaving company sergeant Nelson Victor Carter in charge of the remaining 12th. Nelson Carter and his men made a gallant fourth charge. With Nelson Reed and his mates following closely behind they managed to make a small penetration into the enemy's front and second line but it was futile. Surrounded and heavily outnumbered by the well dug-in enemy, no back up, no ammunition, with total carnage and death all around, withdrawal was the only option to save the last of his men.

By 7·30 am, and now in full light with no cover, Nelson Carter and his small group had decided to get back to their lines. Before retreating, Company Sergeant Major Nelson Carter saw one of the German machine gun nests that was causing severe casualties to his men. Only Nelson Carter had any ammunition left in his revolver. With bayonets fixed the small group rushed the nest as the Germans were reloading. Nelson shot two of the gunners as another escaped. He turned the machine gun on the enemy and shouted to his men to get back to their trenches. His outstanding act of bravery allowed some of his men to get back safely.

As Nelson Reed and his mates tried to make their way back to their own lines, the British released more smoke. It drifted across No Man's Land and instead of helping, it made it impossible for the men to find their trenches. The Germans then opened up with another heavy barrage of bombs.

It was in this place of hell, smoke and barbed wire that Nelson Reed and many of his pals from the Sussex regiments died.

Nelson Carter miraculously made it back through the smoke to his trenches. However on seeing the carnage that his fellow soldiers had endured, he jumped back out of the safety of his trench and started to drag wounded soldiers back. With no thought for his own security, he made several spectacular runs into No Man's Land to retrieve his wounded colleagues. After bringing back at least seven men under cover of the smoke, Lieutenant Howard Robinson, Nelson's wounded commander, saw Nelson stand once more out of the trench. As the smoke drifted he was hit by a sniper bullet and fell back, dead, into the trench. Nelson Victor Carter and Nelson John Reed, like hundreds of others, had died within yards of each other, comrades in arms to the end. The battle was over for the blood soaked earth of Riche-bourg. There had been no ground gained from the action.

The first soldiers had gone over the top at just after 3·00am and the fighting was over within a few hours. During that short period, over

1,300 Sussex men were killed, wounded or missing. There was hardly a town or village in Sussex that did not lose their young men. Many call it the day that Sussex died. Nelson Carter was given a posthumous Victoria Cross for his outstanding actions and bravery – he was 29. Over 30 other medals were awarded for the outstanding conduct and bravery shown on that day.

Press reports were ridiculously optimistic and originally boasted of success with few casualties. In fact only 47 casualties were reported in the local paper. As worried parents waited for news, worse was to follow. The first day of July, 1916, heralded the mass slaughter and the beginning of the Battle of the Somme. As 750,000 British and French soldiers climbed out of their trenches and attacked all along the Western Front, the battle of the Somme began. Within five months, and with less than eight miles of land captured, the combined armies of France and Britain had lost over 600,000 men. Douglas Haig, British commander for the Western Front made it clear that every position, every muddy hole, and every trench must be held to the last man! Men failing in their duty would be shot at dawn.

Such was the carnage that it was not until August that Nelson Reed was officially reported missing in action in the *Sussex Daily News*. Just before Christmas 1916 Jack and Emily Reed received the dreaded form B101-82…"Dear Sir it is my painful duty to inform you of the death in action of your son Nelson. J Reed of the 12th Royal Sussex Battalion on the 30th June 1916."

Nelson Reed's body, like Rudyard Kipling's son, John, was never identified. He lies amongst the countless dead that mark the bloody road to freedom. At Pas de Calais there is the Loos Memorial to mark where so many brave souls lost their lives. The casualties were so high amongst the decimated Sussex 11th, 12th and 13th battalions that they were later disbanded.

It had taken nearly six months for Nelson's parents to be officially notified-six months that they had hoped beyond hope that their son lived, perhaps wounded somewhere or captured. When the few remaining soldiers returned to Eastbourne, many as casualties, they camped along the base of the Downs, at the Summerdown and the Old Town 'tent-hospitals'. It was then that Jack and Emily gathered the full story of how their son had died.

Nelson John Reed was just 19 when he gave his life for King and Country in Flanders Fields at Richebourg. A service was held at Christchurch, Seaside, where he and many of his pals are remembered in the chapel. His name is also at the Town Hall memorial, (though miss-spelt).

I told Hillary all I knew about our long lost relative, and drove away from her home thinking how lucky we all are. Our lives had never been touched by war, real war, when countless young men were served up as machinegun fodder, where letters dropped through letterboxes and families' hearts were broken. Nelson Reed was our own forgotten hero who lies within Flanders field where the poppies blow between the crosses row on row.

GOODNIGHT OLD MAN

The Sun donned a cloud for a hat,
And slipped quietly behind the hill.
Good night old man I thought as
His last rays stroked my wax coat.
The farmer's cockerel took a final tour.
Rest well, for tomorrow once again,
You awaken a slumbering world for toil.

Here I am just before the Queen's Jubilee in 2012. My precious Daimler is out for a run and I have bunting up to celebrate, along with the rest of England and the world for this amazing event. There has probably only been a handful of times in my life when the whole country has joined in a common celebration of joy, and this was one of them.

LEFT: *My handsome half-Russian, half-French father, Igor. He was the most European man I have ever come across, spoke several languages and would prefer to play chess and drink coffee rather than do just about anything else. As he got older all his languages seemed to blend into one, and hearing him trying to order a meal in a Spanish restaurant would make me die laughing.*

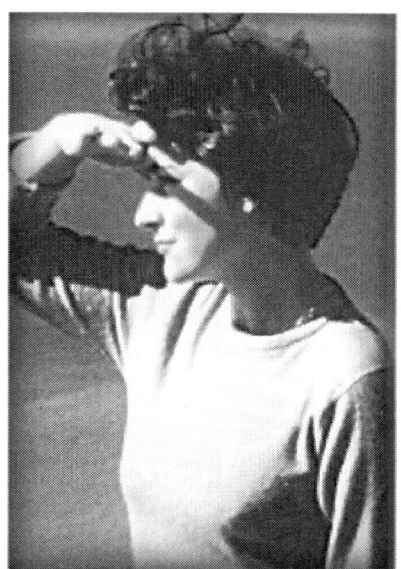

RIGHT: *My half-English, half-Austrian mum, Rosemarie. Blindingly beautiful and intelligent, she would drop into several different languages at the drop of a hat and wind you up in knots during an argument.*

LEFT: *This is the fashion Boutique that my parents and Grandmother ran in Vienna just after the war. My Dad's Russian name helped business, but this was not good during the post-war period and the Russian officers were more trouble than they were worth, often having dresses made for their mistresses.*

BELOW: *This is the cutting room at Simplantex, the family business where I grew up. Lou Moore often joked that she changed my nappy on the cutting room tables. In a manufacturing business this is where it all starts, the cutting of the cloth.*

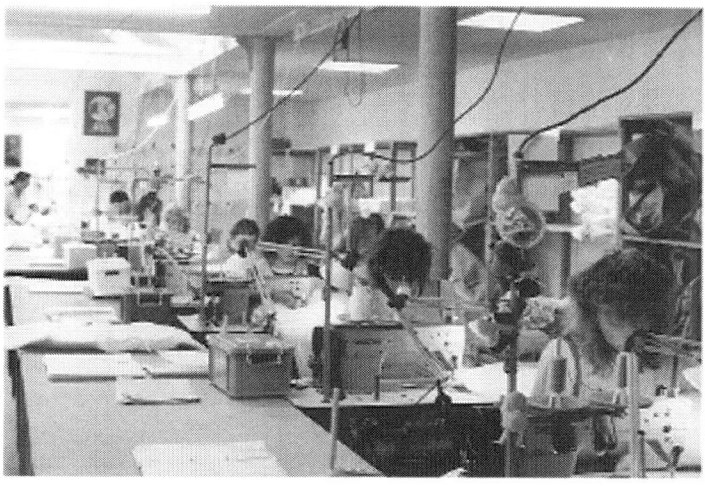

This is one of the sewing lines, paid by piece work with bonuses for beating weekly targets. Normally the shelves were crammed with baby-goods, but the photographer insisted that the picture would look better with no work which was a shame, as I used to love the organized chaos of the sewing room.

The packing department was the final stop before postage all around the world. Lorries came twice a day to collect the goods that went from Iceland to Australia.

LEFT: *Here are a few of the workers at the factory. We ran mainly on outworkers, but there was an essential core of workers at the business. One Christmas, they all made Santa costumes and ran to the restaurant, creating havoc as they went.*

ABOVE: *A small part of the Simplantex range – at one time was the largest manufacturer of baby goods in Europe.*

LEFT: *Here is our good friend Dominic, showing us one of the smallest churches in England, just down the road from Eastbourne, at Chapel Hill in Lullington.*

RIGHT: *This is Seaford Head, just on the tip of Seaford. It is a wild place where only the call of the seagulls can be heard above the waves.*

My beloved Rolly, simply the best friend you could ever have. She was gorgeous and I think she knew it. She made it to 15 before she left us, and took a little bit of my heart with her.

These are the Coastguard Cottages at the mouth of the Cuckmere River. Cottages like these were placed all along the coast to guard against prolific smuggling.

In my first book, there is a picture of my son Tom as a young boy, and here he is on his wedding day in 2012, marrying Corrinne. They must have made someone happy upstairs as it was the warmest spring day for 100 years. Corrinne has spent endless hours correcting my grammar and spelling, for which I can't thank her enough.

This is the annual car show on Western Lawns, Eastbourne, with the Grand Hotel in the background. When the weather is right, thousands of visitors turn up to see the 'old bangers' in all their glory.

WINTER CALLS

Winter was nearly upon us: a bitter, dark morning as black as a crow's wing greeted me when I opened my front door. I pulled my collar up and ran to my car. After a few rumbles, my old Land Rover gurgled into life. I was backing out of the drive when Yana tapped on the window. "Your headlight is half full of water!" I got out and looked at my light, which appeared more like a goldfish bowl. The torrential rain that had fallen from heaven day after day was taking its toll on the old girl. "Look, the light still works!" I laughed and I jumped back into the car, expecting it to go bang any second. "I'll try and seal it up this afternoon when I get back. I haven't got too much work, so I should be back by lunchtime." We kissed through my open window. "Watch your step with that dodgy tyre as well, you must get it replaced." Easier said than done I thought, as I drove away. That would be at least another £150 minimum and, although it was not illegal, it was in poor condition, and with winter nearly upon us it needed to be changed as soon as I could afford it.

I put my brakes on at the bottom of Huggetts Lane and a freezing cascade of water came from nowhere, pouring over my back and down my neck. Like a tortoise I jerked my head down to try and stop more of the awful stuff from going down my back. Somehow the sun roof was leaking and now I was soaked. I put the heater on full blast, but that just froze my feet as the engine was still cold.

Another winter's day had started I thought, as I pulled into the main drag of the London Road and headed for my first call in Horam. It was a simple drop-off. I had repaired the customer's machine over the weekend and arranged to pop it through her door. It was still dark as I made my way back over her gravel drive and hopped into my lovely warm car. The headlight bulb had exploded half way along the A22 but the dull grey morning had woken with a drab winter light, so I didn't need the lights now. As I drove, a stink filled the cabin. At first I thought the car was overheating, but then realised that the smell was far worse. I had trodden in something foul! I had to stop, I couldn't breathe.

I leaned up against the Land Rover as cars flew past towards Heathfield, all ignoring the 30MPH sign. It was something nasty alright. I grabbed my water bottle and gingerly cleaned and scrapped my shoe. A man and his dog marched past. He probably knew that I had trodden in something that some dog had left. He avoided my stare. I also washed off the car mat as best as I could and rolled it up, putting it in the boot to minimise any further smell. I made a mental note to scrub it later.

Once more I was on my way. As I drove, for just the briefest moment, a gap in the sky allowed the deep orange sun to break through. It was as if someone had shone a torch in a dull room. Suddenly the trees, still full of glorious colours, burst out in golds, reds, browns and yellows. The black tarmac glistened and, along with the trees, my view was transformed into a curtain of beauty. Just as suddenly as the magical picture had appeared, it was gone. As it turned out, that was the only sun I would see that day.

"Still got that piece of s##t then?" Shouted Phil as I walked up to his workshop. "Wouldn't give you an old penny for that lump of c##p."

"Don't talk about my old girl like that! If she hears you she may get upset."

"Bah, used to drive one of those. Once we were on a night skirmish in the desert and the bleedin' radiator blew up. We were miles from nowhere. I cursed it all the way back to camp."

"My baby loves me," I said. "One day when I see you stuck in the mud I'll just slow down and wave as I pass." Phil laughed and opened the door to his workshop. I glanced up at his back garden, which he had completely enclosed like an enormous cage. Inside the compound there were seven huskies looking at me as if I was their breakfast.

In the fluorescent glare of workshop, Phil's face looked like he had been in a bar fight. "What happened to you?" I asked, nodding towards his face.

"Oh, I was racing the husky team through the woods and the sledge hit a stump. The sledge bounced up and threw me straight into the trunk of a tree. I was doing about 25. By the time I knew what was what, I found myself bleeding and my team of dogs, faithful to the end, had just run off and left me! When I eventually found them, I could hardly walk and the sledge was smashed up. I'll have to put a new set of Magura disc brakes on it to get it working again."

I went through Phil's industrials, which he used for his harness work and left with him throwing a few more foul remarks at my lovely Land Rover. He also departed at the same time to pick up fresh meat for his dogs from the Burwash Butchers. One of his dogs was 17 and had never eaten dog food in its life, just meat. I drove with the windows open to try and rid the car of its smell which was still lingering like an unwanted guest. My back was wet but at least now it was warm and, wearing navy, as I always do for work, it did not show.

"Morning Dorothy," I said as I walked into the Village Hall along Burwash High Street.

"Morning Alex. I only have two machines for you I was hoping that the group would bring more in but I only have two today. You don't mind do you?"

"Not at all, it just means I'll have a nice easy day." Secretly I thought that I could do with the money, but such is life. I unpacked all my tools and got to work on the first machine. I looked out of the window to see a heavy drizzle set in and remembered my old friend Pearl and her saying, "Nothing will soak you faster than lazy rain. Too lazy to go round you drizzle is, so it just goes straight through you."

Just as I had finished the second sewing machine and closed my toolbox, another woman turned up with a new computer Elna. So I unpacked once more and got back to work. Then another machine turned up, then another. And so the hours went. It was one of those glory days that happens once in a blue moon. More and more sewing machines turned up with assorted problems, and each time I fixed one I was paid. The morning went on in a similar manner. Dorothy kept me well watered with sweet tea and I just stuck at it. By one o'clock I had sorted 11 sewing machines and had a pocket full of cheques and 30 pounds in cash. What a morning. It was the best morning's work I could remember. "Haven't we done well," said Dorothy, as she cleared away the tables that the women had been working their quilts on.

"Best morning's work for years," I said smiling and closing up my tool box. By 1·30pm I was back in the Land Rover, starving and desperately needing a pee. The last time I had seen a loo was 6·30am. I parked in the car park next to the Bear Inn in Burwash and rushed for the public loos. Refreshed and back in the car I looked at my work. It had all gone pear-shaped but in a good way. I only had one more call, which was supposed to have been late morning at Jean Jarvis's. I pointed the car towards her at Punnetts Town and rolled down the filthy roads with my wipers squeaking as they smeared the greasy drizzle over my windscreen.

It was just a grimy day, and by two o'clock it was already showing signs of giving up. I passed the old graveyard and chapel at Chapel Cross on the main road. Mist was making its way up the valley and over the road towards the graveyard, creeping and circling as it went, cold fingers of vapour clutched at the gravestones and a crow, perched on one that was leaning precariously, gave out three load squawks. I shuddered and drove on.

About half way down the long muddy track to Jeans, a thought suddenly hit me: was it her I was supposed to be calling on? I only exchanged the briefest of words with Yana earlier and it suddenly struck me that I might be going to the wrong customer! Jean lives well off the beaten track down the smallest little twitten between two fields.

I wobbled gingerly down with my tool box rocking in one hand and my work details and kneeling cushion clutched tightly under the other arm. It was difficult going as the path had become a quagmire and as slippery as ice. I hopped from one side to the other trying not to step into the deepest mud. The drizzle made it almost impossible to look up as it immediately got in your eyes. The panoramic views down to the sea that normally pamper the eyes along her path were gone. The hedge line was just visible and a miserable flock of sheep huddled in the far corner, pointing their bums at the prevailing wind. As I neared Jean's cottage the cold was seeping into my bones; I felt weak and tired. Concentrating on so many machines without a break was probably like a surgeon who has to stick with a tricky operation hour after hour. The big difference being no one was going to drop dead if I messed it up.

I opened Jean's creaky old gate and looked around. Even if this was the wrong call it was too late now. Frozen, starving, wet and exhausted, her little cottage looked like shelter. I felt like a pilgrim nearing a cathedral at the end of his journey. Glancing at my watch I could see I was over two hours late, something that I never am as I am a stickler for time keeping. Even if I am a few minutes late I'll usually phone ahead, but today I had no phone and a whole string of incidents from early dawn that had conspired to make me run late.

I peered into the window next to the back door. Inside was a perfect scene and brought a feeling of comfort. In the warm light I could see a large wooden table and on it was a Singer 730 with a few bits taken off. I knew I was in the right place. Between the wooden table and the Aga cooker was Jean, sitting in her comfy chair with her cat on her lap. She was reading to him and stroking him at the same time. It always amazed me how independent she was in her little cottage way out in the middle of nowhere. I tapped on the window and she looked up with a smile. Oh how sweet the room was on that dark wet afternoon. Jean immediately put the kettle on and I hung up my wet jacket near the stove.

"Good heavens, your stomach is rumbling worse than Fred's tractor. When was the last time you had something to eat?"

"I had a couple of pieces of toast just after six this morning," I said, with my most pitiful, famished, voice. It worked like a dream, I didn't even have to fake passing out from fatigue!

Before long I was sitting in front of a plate of cheese and crackers with a hot coffee clutched in my cold hands. "I am sorry I am so late Jean, it was just one of those days from start to finish."

"We all have them Alex, and I knew that if you were not coming you would have let me know. You haven't let me down for the last 25 years so I was pretty sure that you would not let me down today."

As the afternoon rolled on I felt my body revive. I fixed her machine and chatted about old times, and by 3·30pm I was once more making my way up the muddy path. Half way up, and breathing hard, I stopped to look at the miserable scene. Dark was nearly here and miles of open views were being stolen by the oncoming night. Near the top was the welcome sight of my old girl waiting patiently for me. I packed my tools neatly away, jumped in and fired up the beast. With the heater on full blast I rolled back down towards Eastbourne.

As I turned into Church Street I was squinting from the oncoming headlights, another day was nearly over for me. I pulled into the drive where I could see Yana, who I knew was making her new pumpkin and poppy seed chutney; she waved from the kitchen window. What a day, I thought as I closed the car door, threw my dirty rubber mat out into the rain, and ran for the front door. "Get the hot water on darling, I need a bath and some hot food."

A huge cauldron of chutney was bubbling away making the kitchen windows steam up. On the sideboard was an assortment of freshly washed jam jars waiting to be filled. "What about the light on the land Rover?" asked Yana. I wiped the window and peered out into the cold darkness.

To her amazement, I pulled out of my pocket a wad of cheques and cash and laid it all on the table. "Sod that," I said with a huge smile. "I've earned enough money to buy a new light and a new tyre. And even better enough to go out for dinner as well. What do you fancy, Italian or Chinese?"

DAYS OF OUR LIVES

What a great day, I was thinking as I pulled out of Cotchford Lane, near Hartfield into Jib Jacks Hill. All my calls had gone well. I had a few cheques in my pocket and enough cash to put fuel in the old Land Rover. I drove past Hundred Acre Wood where Winne-the-Pooh and many of his friends had come to life. I cut across Ashdown Forest and made my way through Crowborough, towards Rotherfield: I made a mental note to keep an eye out for Lisa Marie Presley now that she was living in the vicinity. Elvis wasn't exactly working down the chip shop but amazingly his daughter, Lisa Marie, was helping out in a mobile chip van and singing at the King's Arms.

My call had just been at a home near Cotchford Farmhouse where A A Milne had written his immortal stories for his son Christopher Robin. The farmhouse was bought by Brian Jones from the Rolling Stones, who had died in mysterious circumstances in the swimming pool. At the time the press had a field day with his death and there were many dark theories and speculation, but no one really knew the truth. Locals say that he had an asthma attack while swimming and drowned. As sad and as simple as that.

From Rotherfield I took the back lanes across country, down past Lovers Lane and into Battle. I always wondered how that lane got its name and pondered if I would ever find out.

At Battle Abbey the Normans were attacking again, this time from several coach loads of students disembarking French coaches to see where their most famous battle took place. One more call in Crazy Lane next to the Pestolozzi International Village near Sedlescombe, where rumour has it that some of the stolen loot from the Great Train Robbery is still buried. Then down to my call in Silverhill at the top of St Leonards.

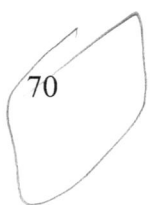

I was sitting in a close, writing down notes from my customers' machines and putting them all in order before my last call of the day in Bexhill. Suddenly the car door opened and a white-haired old girl hoisted herself up onto the car passenger seat. I looked at her quizzically wondering what on earth was going on. "You're not my son!" She said staring straight at me.

"No I'm not. And you should be in a home!"

"I am, it's my day out today. Where is Tony?"

"How should I know?"

"You're a very rude young man."

"Well you've just climbed into my car!" I laughed.

"Oh, I suppose I have," she smiled back and giggled like a girl. "Never mind, I am sure Tony will be along any moment. You know you look just like my husband. I first saw him at a dance in Worthing. It was love at first sight; I went all weak at the knees. He was in his Navy uniform and I skipped across the dance floor and interrupted his dance. It was an 'Excuse Me' dance so I was allowed to barge in. We danced all night and I didn't let go of him for 50 years. He's gone now. Oh look, there's Tony. Bye-bye. You take care."

With that she hopped out of the car and tottered over to Tony (I supposed it was him), who was looking at me with an apologetic smile as if he knew what had happened. Only in Silverhill, I thought, waving to them as they passed. I smiled to myself and headed towards Bexhill and my last call.

As I was driving I looked up towards Lavant Close. I pulled over for a moment. Sid Day and his wife Eve live there and Sid was what the Londoners call a 'real diamond'. Born in Kennington, within ear shot of the famous Bow Bells of St Mary-Le-Bow, he was a proper Cockney. He once told me how when he was on the trucks doing deliveries, they were taught how to stop their vehicles after the first rain. Today it is not a problem but when London was still full of horse-drawn vehicles their droppings dried on the black pitch roads. When the rain came two things happened, London, like I suppose all large towns and cities of the time, stank, and secondly everything turned very slippery. Apparently wet horse muck is the closest thing to ice on the road. Sid was taught how to slide his truck into the kerb and use the kerb to slow his vehicle.

I remembered the last time I called on him. If ever there was a man full of stories it was Sid.

"Well if it ain't Fanny Fernackerpan with his tool kit." Sid laughed as he opened the door to me. "Come in son. Kettle's on and Eve is dying to see you; her machine is driving her crazy. Your name's mud, I'm telling you. Got your cricket box on?" Sid patted me on the back, laughing at his own jokes as he showed me into their neat living room. "I'll just get 'er indoors."

I didn't need a cricket box, Eve was delighted to see me and fussed around me, making sure that I had everything that I needed so that I could get down to work on her old Brother sewing machine.

"How are you doing Sid?"

"Can't complain. Life ain't no bed of roses; no pleasure cruise."

"Sid," I interrupted. "Don't you go breaking into song again. You know it scares the cats!"

"I sing like a bird I do, sweet as roses in June. Anyway you asked. I ain't going nowhere fast now as me bloomin knees are shot. I'm on 18 pills-a-day to keep me mobile and I ain't 'avin no operation, 'less they drag me there kicking and screaming. I mean I'm not far off 90 now, so I ain't expecting no glory days in front of me. I pop the pills as soon as I wake up then roast my knees with the heat lamp. Then I put frozen peas on me 'hunnies' and by that time the pills have kicked in and I can move."

I looked quizzically at Sid as he pointed to his knees. He had slipped into his old cockney slang for knees, Honey bees...knees. Having a conversation with Sid was always a laugh, as I tried to keep up with his unique ability to completely confuse me.

"You wouldn't want to walk a mile in my shoes, son," grunted Sid as he wobbled off and started searching through a pile of papers. I sat down at the heavy old 1960's Brother sewing machine and got to work.

Sid and Eve had been together forever and it would not be long before they celebrated 60 years together. The first time I met Sid he regaled me with stories from the old days; from the war and other memorable tales. Although he was a Londoner and his parents lived there, his favourite story was about his Uncle Albert who lived in Eastbourne.

Albert Blake lived in a terraced house along Hurst Lane and worked on the railways. He had moved down from London in the 1930's and after a stint selling whelks and winkles from a wheelbarrow, he managed to get a job on the railways. He started as a ticket collector and worked his way up to porter, then guard on what he used to call his 'cloud makers'. Everyone took Albert as a happy-go-lucky, mild-mannered man, but he was the stuff of heroes, and one day he would prove it. Albert had served in the Army stationed in India as a boy and by the time World War Two broke out he was well past active service and carrying on his duties on the railway.

When Sid had leave from the army he had a choice of popping back to his parents in London or going to Eastbourne to stay with Uncle Albert. Living behind the Elephant & Castle in London Sid's upbringing was rough'n'tough. The choice was a dirty bomb-torn inner city or a clean seaside town. It was not a hard decision to make. When you booked your pass you only had one destination option for your leave, so Sid often chose to stay with Uncle Albert. Funnily though, nearly all the trains had to pass through London so Sid often managed to get to see his parents and Albert with his one travel permit.

Albert was a great piano player and his local was the Hurst Arms, on the corner of Mill Road and Willingdon Road. The picturesque town pub fed and watered the locals who lived in the rows of terraced houses all around. Opposite the pub was a newsagents where, years later, I would pick up my parent's cigarettes, even though I could hardly reach up to the counter. Back in the 1960's regulations were a little lax, to say the least.

When Sid arrived on leave in Eastbourne, every night was party night. He would scrub-up and rush to meet his uncle at the pub. He would make his way through the black, unlit streets and follow the sounds, feeling his way to the door. He would carefully pull back the black-out curtains and move inside, closing them behind him before opening the inner door. Inside, the pub would be in full swing. Through the smoke Sid would see Albert, who would be banging out a tune on the old piano with locals and many Forces personnel singing along. Beer would be flowing and Sid would push through the crowd and join Albert on the opposite side of the piano. Between them they would knock out tunes to the crowd, who in turn would help with drinks. Their 'shin-dig', as Sid called it, would go on all evening and if Sid spotted any girls he took a liking too, he would smooth-talk them as he played away at the piano often sharing his drinks.

Billeted nearby were young women from the WRNS (Women's Royal Naval Service), who became known as 'Jenny Wrens'. Many were studying radio training and deciphering at St Bede's School along the seafront. Eastbourne College which had joined with Radley College had closed as a school, but part of the school buildings were occupied by the Royal Navy Torpedo School after their Portsmouth buildings were destroyed by bombing. Wrens and WAAFs would often travel from their surrounding billets at Newhaven and Brighton to Eastbourne for some fun. Eastbourne was full of young servicemen out for a good time. During the later part of the war, after the bombing raids tapered out, the town was far more peaceful.

Now and again Sid found that some of the girls were up for a laugh. Occasionally, when the weather and company was right, Sid would chat up some of the 'Forces Sweeties' and take them swimming, a rare luxury during the war. The girls would not have costumes but that did not stop them skinny dipping.

This was not as easy as you might think, for although Eastbourne was a seaside town, the beaches were strictly off-limits. The beaches were mined and wired. Even the pier had a huge chunk cut out of it in case the enemy tried to moor a ship alongside it during an invasion. At the Redoubt Fortress, along Royal Parade, and all the way along to the slopes of the Wish Tower there were armed troops manning anti-aircraft guns. The War Department had requisitioned, amongst others, the Chatsworth Hotel which was occupied by the troops who were manning the coastal defences. The Home Guard also patrolled the beaches all day and night. However all this was not going to stop Sid and the Wrens from their skinny dipping.

They would cycle, with their towels, down to the railway depot through the pitch dark streets. Cycling at night was not that easy for there was a complete blackout in force. If a light shone you could be fined; 30 shillings for a minor infringement or £2 for a more serious breech. That was two weeks wages! Even if you were lucky enough to find some batteries for your bike lights, using them was a gamble. The Hurst Arms was on a hill so the trip down to the station in the dark was fast and exciting. At the depot, Sid would borrow wire cutters and pliers from the works stores. Then they would cycle along the back streets and park up in Cambridge Road. They would peek out along the road to make sure the Home Guard were nowhere to be seen and skip across to the promenade. In the dead of night they would creep to the seashore. Sid would make short work of the barbed wire. Being local, Albert had told Sid exactly which parts of the beach were mined. There was patch of sand near the Martello Tower that was not mined and they would silently slip onto the beach.

There was nowhere near as much shingle in those days, and sand came almost up to the sea defence wall of the promenade. Under the summer moon they would strip-off and skinny dip in the fresh sea water. Giggling was kept to a minimum!

Later they would dry off and then carefully close up the barbed wire. From a distance no one would know that they had ever been there and the tide would steal away their footprints. (It was important to seal up the wire because, if a hole was reported, everyone would be on high alert for spies). Then a quick stop next door to the Salvation Army where they used to serve hot tea all night. They would then return to the railway station where they would drop off the wire cutters and pliers and meet up with Uncle Albert on the early shift. Here they would have a 'boiler's breakfast'. Albert would fry up some bacon and eggs from the station supplies on an old oiled shovel heated in one of the train boilers. Then finally back up the hill to get the girls home before they had to start their day's training.

Sid probably caused a few low scores on exam days! As Sid told me more than once, they were the best days of his young life.

"Here it is me old mate, the certificate I have been looking for." Sid leaned over the table and offered me the picture. In the gold frame, which was about 10 inches square was a certificate bordered in a floral pattern. Inside, written in bold calligraphy were the words...The Chairman and Directors of the Southern Railway wish to express to Albert Henry Blake their appreciation of the courage & devotion to duty displayed by him. 1943

"See, I told you he was a hero. There is also a plaque in Eastbourne Town Hall somewhere but I ain't never seen it."

"Sid, that's amazing. Please, please tell me again what happened." I sat back with a cup of tea and let Sid get into his stride. His eyes sparkled and lips turned up into a smile, and 70 years disappeared from his face. All of a sudden I was looking at the young soldier, so proud of his uncle.

"It was like this, see, now I wasn't there but I did hear it from so many locals – Albert hardly had to buy a drink himself for years. He was on normal guard duty down at the terminus of the railway station when the air raid sirens howled. Everybody ran for the shelters but Uncle Albert saw some planes had spotted the smoke coming from the railway station. As Albert ran with the rest of them he glanced back at the long ammunition train that had just pulled into the station: the train driver was nowhere to be seen. The planes turned over the hills above Eastbourne to come back to the terminus! It was obvious to Albert what their target was. If they hit the train, half of Eastbourne Town Centre would disappear, as well as hundreds of lives. Instead of running for shelter he ran for the train. He slammed the engine into reverse and managed to get the beast moving as the planes lined up on it. Now, the Eastbourne railway line curves as it leaves the station, and although he was no train driver, he got a good head of steam up as the planes opened fire. They hit some of track and bits of the train but luckily found it impossible to hit the ammunition as Albert curved out of town under the bridges."

"Our pom-pom guns were going mad, firing ack-ack and antiaircraft missiles at the aircraft. After two runs the planes turned tail and disappeared out to sea. Albert stopped the train and brought it back to the station, where the ammunition was later unloaded and safely stored away. He was a hero, a real bona fide hero. Nobody could believe that the quiet train guard could have done something so amazing. With no second thought for his own safety he had risked his life for the town. He was awarded a special medal, which the mayor later presented to him. He was all over the newspapers. It was all quite embarrassing for Uncle Albert, but he did appreciate the free drinks. I tell you boy, those times were the days of our lives. Now 'ave you finished the machine or do I 'ave to pour you another cuppa?"

That was Sid all over: one for a great story and a life full of optimism. Even though he was in constant pain and had a dozen medical problems, he was still going. He was even putting the finishing touches to a boat of his, which, as he put it "God willing," he was going to sail up and down the River Medway.

After the War, Albert settled down and loved his daily routine, he would set out to work each day on his bicycle and ride down St Annes Road to the back of the railway station. His trains always ran on time. As the day wore on and the clock struck four, Albert's wife would come out to their small back garden, walk down the narrow brick path to the vegetable patch and stand on an old chair. In her hand she would hold a broom with a duster tied on the end of it. As the afternoon train passed in the distance she would wave to Albert. Albert would be waiting in the open carriage door as his cloud maker rolled by. As soon as he saw her he would wave a flag back. If Albert waved back with a white flag it meant that he had manage to get some supper and would bring it home to cook. If he waved with a yellow flag it meant that she should meet him in town, by the station clock, for fish and chips. They carried on their signalling for many years, (much to the amusement of her neighbours).

Albert worked for years on the railways, and when he eventually retired he managed to get a part-time job helping at the Towner Museum in Gildredge Park. He was good friends with Tommy Cooper and they would often drink together at The Star.

I started the engine and pulled into the traffic, making a mental note to call on Sid and see how his boat was coming on. He did promise me I could be his shipmate on his maiden voyage up the Medway.

After my last call in Eastbourne, I drove along the main road near where Sid and Albert used to have so much fun. The Hurst Arms still looks much the same as it must have done during the War. If only those walls could talk, what tales they would tell! I could only imagine the laughter slipping out of the cracks in the black out curtains all those years ago, as Sid and Albert banged out tune after tune on opposite sides of the 'Old Joanna'. Sid, with his devilish smile, chatting up the 'Forces Girls' who would be singing their little hearts out. I can see Sid leaning over to them, "Fancy a little skinny dipping girls?"

Today all the trees and gardens are overgrown along Hurst Road and you can only see glimpses of the trains as they thread their way through the suburbs to the station terminus. However, a long time ago, a local hero often waved to his wife, who was balancing on an old chair in the vegetable patch, as his cloud maker went by. They were, As Sid said, the best days of their lives.

Sir Marmaduke Chatfield

Sir Marmaduke Chatfield has bit the dust,
And all his medals are turning to rust.
So much bluster and so much blow,
Was little more than pompous show.
Now the worms feed upon his bones,
And he dreams no more of kings and thrones.

THE REAPER COMETH

No leaf on tree, no bird in song, just a black bitter wind throwing daggers from the east.

It was dark and cold. I had dropped Yana off at work and parked to walk up to my usual little paper shop along Upper Carlisle Road, called Carlisle Newsagents. In the darkness everything felt a little unsafe, dangerous. Eerie shadows were being cast from the street lamps and the wind was whispering in the alleyways. The rows of shops and cafés that normally looked so appealing with customers buzzing in and out were empty, hidden behind the black glass.

I became aware of footsteps behind me, coming up fast. I quickened my step, but they were still approaching. I could see the welcoming light of the paper shop but I instinctively knew it was too far. Suddenly a shape was by my side. It was a small black dog followed by its owner on an extending lead. It sniffed my trousers and moved on. The man strode past in the darkness with a stride that would make an Olympic walker proud. I subconsciously breathed a sigh of relief.

I passed the usual pleasantries with Alan in the shop. We were pleased that the world had not ended with the Mayan Calendar, as quite a number of people had predicted, and instead it was the usual hectic run up to Christmas. Alan, even at 84, was always upbeat and positive and had placed his orders for the seafront hotel papers, just in case the world survived and he had to come into work!

I walked up to the seafront as the first seagulls made their morning circles of Eastbourne, searching for scraps from high up in the early morning sky. It was always astonishing to me how they could possibly see anything in the dark streets below. However they knew there would be everything down here from discarded burgers to old chips dropped from the night before. Seagulls, especially herring gulls, are the unsung cleaners of all seaside towns. They convert our waste into fertilizer which they then spread without prejudice evenly among the flowerbeds and cars of Eastbourne.

I stopped at the railings on the seafront to peer at the horizon. It was still dull and grey with only the faintest hint of dawn. It was low-tide and the man who had passed me earlier was throwing a ball to his little dog. They were being reflected on the wet sand and in the graphite light they looked like a charcoal drawing. I suddenly felt a wave of emotion flood over me, the man was doing exactly what I had done so many times with my beautiful little Patterdale, Rolly. Rolly had lasted 15 years but in that terrible imbalance of nature she had lasted no longer, and left me with a chunk missing from my heart. I had carried her little dead body one last time around our garden, which she had patrolled all those years, and later scattered her ashes in all her favourite spots; where she sat in the sun, where she lapped water from the pump by the pond, where she barked at people walking down the alley besides our house. God, how I cried.

Death had been an unwelcome guest in 2012. He had arrived in January on an icy wind with the sudden death of Yana's father. A desperate rush across a frozen England to York was followed by our crunching footsteps through the snow to the Pontefract Crematorium. Clive came and went to the sound of Glenn Miller. He would have approved. Both his sisters soon followed, and as the year progressed, one-by-one the Grim Reaper took my friends and family. Parsley, the last of my eight cats, meaowed her way to sleep and I buried her with her twin sister.

Sometimes Death was only hours ahead of me. I called on Ellen Tompkins for her usual service. The week before she had sounded her normal bubbly self. I had managed to get hold of some dolls' patterns that she used to love making, and I knew she would be delighted. I had known Ellen since she sewed for my parents at the family firm. There was no answer at the door and none of the usual barking from her small dogs. I walked across to Hazel Mitcher and knocked on her door. "You're too late Alex, they took her away – she died this morning."

Lin Hall, my old editor, who had tried so hard, and failed miserably, to get me to write proper English took his last breath. All year Death stole them from me, sweeping his scythe across my life with little regard. In the summer I cancelled my yearly jolly to the local air shows as my old friend Cyril Johnson would no longer be with me, laughing at all my silly jokes and posing with the Utterly Butterly girls or Red Arrows pilots. A pain in his hip turned into the dreaded C word and within weeks he was gone. His smiling face on his scooter as he buzzed into my drive was now a memory that made me smile and well-up in one go.

I often called on one of my customers in Hampden Park; her husband Stan was seriously ill and had been for a long time. I used to look at him in his special chair in the middle of the sitting room. Around him was an assortment of medical equipment, oxygen bottles, pipes and tubes. Amazingly his favourite programmes, which Sophie his wife put on for him, were hospital reality programs. As if he did not have enough of it in real life. I once asked him why? "I know it all," said Stan. "I am as familiar with the inside of a hospital as any doctor. Since my lungs were shot I'm in those places as much as here. I still roll out of them though so they ain't got me yet." Stan grabbed his mask as he started to laugh and needed to suck up more oxygen. Sophie used to sew in the sitting room so that she could be with him, and he had great hearing. As soon as the sewing machine got a bit noisy he would tell her to ring me to get it serviced.

This summer I called, and as Sophie opened the sitting room door for me, instead of being greeted by old Stan with his usual sarky remarks, there was just his empty chair. I said nothing but I instantly knew he was not just on one of his trips to hospital. The room had been cleared and the only evidence of all his medical equipment was one small oxygen cylinder left in the corner by the door. I stopped dead in my tracks as if nailed to the carpet. "Yes he's gone," said Sophie with resignation. "He put up a good fight. When the specialists first diagnosed him they gave him three to four years and he lasted nine. Proved them all wrong for a while, did my Stan, love him. You'll find the sewing machine a bit noisier than usual as I didn't have my old man to tell me it needed sorting."

And so it went on. My old mate and neighbour Norman had gone to one of his great ships in the sky. His antics during the war as a naval commander had been the subject of one of my books, *Norman, a Journey Through Time*. I remember so clearly his upright little figure marching up and down Church Street. As he aged, his trips got slower and his shoulders dropped until just posting a letter took an age, but he kept at it. Each night he would pour himself a 'drop of grog', stand to attention and salute the world with the same words he saluted his crew each day during WW2, "Up spirits and stand fast for the Holy Ghost". Then down it in one. It made me smile to think of the endless hours I had spent writing down his life story, only to have lines crossed out and corrections made. Norman was a retired school teacher and he marked each chapter of our book like one of his pupils' homework. His little eyes would sparkle as he often toddled off, and then he would return with an old buckle or in one case, a samurai sword. "Took that off a Jap officer back in '45. He didn't put up much of a fight! That lump came from a German soldier in St Peter Port, where we swept the harbour for mines."

When they carried his small coffin into St Mary's Church in Willingdon I could hardly breathe, how could such a great man ever die. Even the Bishop who had come in especially to oversee our friend's final performance choked up, and had to gather his faltering words before continuing.

On and on it went, bad, bad, bad. By December, Death had got fed up and looked elsewhere for new stock. His departure brought a welcome respite and there were surprising escapes.

Violet Perrin had been unwell and I feared the worst, but she somehow slipped by the Reaper once again. I had even let her keep my favourite body warmer to keep her cosy. Perhaps that was what saved her! The Reaper did not recognise her tiny little frame hidden away inside my huge padded warmer. She was my oldest customer, and at 97 was still independent. Born during the First World War she had survived everything that Death had thrown at her. On her old cottage wall there is a picture of her taken in 1934 when she was 18 and stunning. "Jerome & Jerome took that. I seem to remember that they were in Western Road in Brighton and the best photographers in the business at the time. I can tell you some stories Alex" she used to giggle, as she watched over my shoulder while I fixed her sewing machine. Funnily, she was often so close that she actually leaned on me for support. Being hard of hearing, the closer she got to me, the easier it was for her to hear. "I was working at the 'Baccy's' in the Brighton Railway Station through much of World War Two. Evacuees, lost loves, forces romances, you see it all at a busy station." Vi was a gem and so sharp she could slice fresh bread with her tongue. Yes, at least Vi was one of my survivors.

Wendy Lloyd, cheated death as well. A regular sewer, I would pop in now and again to sort out her sewing machine and she would enthral me with old tales of Eastbourne. A violent reaction to antibiotics left her in emergency. The week had started much the same as any other for her but treatment for a simple infection left her dying. She drifted in and out of consciousness as nurses and medical staff rushed around her. She heard the emergency specialist Dr Patel powerfully instructing the theatre, "she has less than two minutes where's that adrenaline?" Wendy felt the searing pain as a syringe pumped life saving chemicals straight into her heart. Wendy survived to fight another day.

I had no comprehension of the pain that I would feel this year, no knowledge of how death sucks the life out of you little by little. How you plod on with your insides burning. How the smiles would not come so easily or last so long, how a familiar face would send emotions and memories surging through my assaulted body. They say that time heals all wounds, but the cold reality of death is that loss can never truly heal because I never want to forget.

I gripped the cold blue rail in front of me as I felt my legs weaken and lifted my head to stop any more tears. I was vaguely aware of a faint electrical hum behind me but did not bother to look around. "Hard isn't it?"

"Sorry, what?" I spluttered.

"Hard losing a dog."

"How...?"

"Oh, I'm not so green as cabbage-looking. I used to see you walking along here. I live in the flats opposite," said the old dear in her wheelchair. "She was a lovely little dog that you had. What was her name?"

"Rolly," I answered, clearing my throat a little.

"Unusual shape, what was she?"

"She was a Patterdale. I picked her up from a farm that I had visited. Come to think of it I didn't even want her, it was a mistake really."

"Well she seemed like a happy little mistake to me," said the old dear humming off in her wheelchair. "The best ones often are," she shouted over her shoulder as an afterthought.

I smiled and thought back to that first meeting with Shelly Froggatt that would change my life. It was a simple sentence that started a roller coaster ride with my mad dog. "If I ever had a dog Shelly, that would be the sort I would get. What is she?"

"She's a hunting dog from the Lake District. They are called Patterdales. Fearless; they keep the vermin down on the farms – as tough as a horseshoe and totally faithful. She will never let you out of her sight and protect you to the death."

That was how our journey started. About a year later the phone rang, "Shelly here. I have your dog ready. I have sold the other six and kept the one that looks just like mum with the short black hair. She is eating me out of house and home so the sooner you pick her up the better. Oh, we have named her Rolly Polly... Eats like a horse."

That was it. I sat in stunned silence. Yana looked at me questioningly. "Shelly Froggatt has our dog!"

"What?"

"Shelly Froggatt has our dog. She is called Rolly."

"We don't have a dog and I would never call one Rolly. What sort of name is that? Did you say that you wanted her dog?"

"No, er, I don't think I did, but I seem to remember saying that I liked her dog. Now she has apparently produced a litter and Shelly has saved us a little fat one. What do we do?"

"Well we could say you are allergic to them, or something like that. We are up her way on Tuesday so we could take a look and see what we think. I don't mind having a dog if you want one. The kids would love it. How big is it?"

"Patterdales are small, bigger than normal terriers like the Jack Russell's but with longer legs. The one I saw was sort of dark, yes, maybe it was black. Imagine a sleek half-sized Labrador, that sort of shape anyway, with short hair."

"Well, we'll have a look and see what happens," said Yana. And that was that.

No sooner than we laid eyes on Rolly, we fell in love. She ran up to me and I picked her up. She lay in my arms like a little baby, trying to lick my face. From that second on I forgot all about fake allergies. Rolly came home with us like an excited child. On her first day home she took over the place. She pissed everywhere, completely ate one of Yana's only pair of Armani high-heeled shoes and tripped me over at least a dozen times. By nightfall we were both exhausted and looked at each other with despair. When we went to bed she howled for hours until, at midnight, we gave in and let her sleep upstairs. It was like having six children. She was just small enough to squeeze out of the cat flap if she stretched her back legs out, something that Patterdales can do. As soon as she was out she patrolled the grounds, somehow figuring out the boundaries and never going beyond them. Even if the gate was open, there was some sort of invisible line that she drew. When she was small I could have sold her a thousand times; she melted the hearts of all who saw her, even the vicar who was thoroughly impressed at they way she killed one of his graveyard crows in front of my gawping mouth! I walked around the graveyard, not through it, for years after that.

And so the years passed. Rolly and I walked the highways and byways of Sussex, tramping endless miles down endless paths. Eventually she seemed to know every back path that I took as if her memory was programmed with a map. My son Tom taught her so many tricks that she would have us all in stitches, entertaining us at the drop of a hat; playing dead or balancing a treat on her head. The bond between us was so strong that it was impossible that it could ever be broken. She never seemed to age and even when she was 12, she looked and acted like a puppy.

But, isn't there always a but! Time caught up with her and as she reached 14 all the aches and pains of old dogs came to visit her. The dog who once could ran faster that we could cycle as we raced through the woods on our bikes, could hardly walk. I would take her out for a little stroll but she was happy to just sit in the sun and watch the days slip by. One day I took her for a walk and she just sat down. No coaxing could make her move, she just looked at me with her deep brown eyes as if to say 'no more'. I carried her home and that was our last walk together. No more exciting trips to Beachy Head where she would hunt crabs in the rock pools, no more chasing other dogs like the wind in Babylon Woods or Butts Brow. Her final year was sitting around by the garage waiting for me to come home from work, her only sign of excitement by then was her wagging tail. She stank and we bathed her every other day, we became her carers running around after her. Exactly what I said we would never do. And then, as suddenly as she had come into our lives, she was gone, and with her a chunk of my heart withered and died.

I could not understand the pain I felt, perhaps it was the accumulation of all the pain that had been thrown my way this past year. There is a bond between humans and dogs that is stronger than anything I had ever imagined. All dogs are descended from wolves. Over thousands of years, dogs and humans have come to know each other and they have a friendship that is unique on this planet of ours. They say that dogs even learnt to bark (something that wolves don't do) to try and communicate with us.

I watched the man and his dog running about in the morning light knowing exactly how he felt. I remembered how one day I had gone fishing with Rolly. It was low tide and to get my bait out I waded out into the surf. At Norman's Bay the water is shallow when the tide is out. I waded out into the sea up to the lip of my chest waders and turned to look at the bait on the end of my rod to make sure it was not tangled before I cast out. To my amazement there was Rolly paddling for her life, she had followed me all the way into the sea. Had I decided to walk to France, she would have come with me. I cast out and laughed all the way back to shore with her little wet body tucked under my arm.

Time to go I thought, and turned for the car. My walk back to my Land Rover on that bitter winter's morning was filled with images of one man's best friend and my mountain of happy memories. Perhaps that was the secret, perhaps that smelly black mutt that I never really wanted, that used to silently fart when we were watching TV and creep out of the room before we started shouting, was teaching me how to deal with loss.

It was a cheering thought that, in the end when all we have are our memories, my special little friend may once more fill my life with laughter.

TINY ACORNS

I hear the birdsong as autumn leaves lay like a chestnut sea,
And gaze with saddened eyes while gentle rain entrances me.
So swift the seasons run this year, so soon the swallows fly,
While church bells toll to summon all for a friend has passed us by.

Oh how the tears do swell with the choir in harmony,
Where angels dwell in every note and where my friend should be.
I dare not glance at other's eyes should I feel their pain in me,
This death has shaken all our souls with a cold reality.

We stout men of England strong that grew from soil so rich,
That beat the anvils from our past and learnt their rhythmic pitch,
We walked the path of duty, to earn our right to stay,
But beneath the churchyard yew is where we grudging lay.

What happened to those sunny days and our endless schemes,
When we were tiny acorns amongst a field of dreams.
Wide eyed we raced through life, our hearts were brimming full,
Yet the final flag is a shroud of silk, so harsh and miserable.

Were I to take your hand my friend, were I to grasp it hard,
I'd gladly share half my days though it would mark my card,
Then through the lofty corridors where silent whispers fly,
We'd shout our voices hoarse my friend and laugh until we die.

CORAL

My car wheels were rolling and the day was running like a well-oiled Swiss watch. I had to smile to myself as I drove along. Today was a special day, my 34th wedding anniversary. After work, Yana and I were going out to celebrate. The sun was out and the birds were singing, the back of winter was broken and spring would be here within a heart beat. All was well with our little world. I made a mental note not to buy a newspaper and get dragged down with all the negative news that often fills the daily pages.

We have the benefit in England of having four distinct seasons, and although we know when they are coming they can still take us by surprise, like the first frost on the car in late autumn or when you see the first daffodils in spring. The British are obsessed by the weather, we include it in almost every conversation, as we huddle like hermits through the long winter months, visiting indoor shopping centres and friends and peering out through double-glazed windows at the world outside. Only the brave few hike and cycle. All that changes when spring approaches. Spring is like the extra birthday that we all get each year, and our excitement builds as those first warm days draw ever closer.

By late February, winter was having her final fling, and as I drove and I could see the daffodil buds swollen, ready to announce that magical time of year. It is funny how the worst and most miserable month is followed by the best month of the year. Soon those buds would be bursting with colour as the finest season of the year explodes. Birds that had quietly shivered all winter would fill the morning air with song. Nothing fills your soul with joy more than the first bright spring day. I knew that our grey old country would shortly be transformed, and months of dark dreary weather would be forgotten in an instant. Women will gossip about the weather over fences and checkout queues more than any election and winter would make her final curtain call. The clouds will part and we will be shown one of the most beautiful places on God's green earth.

The bright day had lifted the spirits of my customers and they were in chatty moods. I had money in my pocket and a string of successful calls under my belt. There is no doubt that any self-employed person gets a buzz from getting paid. It is like a pat on the back, a well done for your work. An added bonus brought on by the sun was that my chatty customers were telling me some great stories.

"That belonged to Cats Eyes Cunningham. Well that's what I was told. No reason for them to lie. It comes from a Gypsy Moth biplane." My customer was standing next to me and we were both staring at a huge wooden propeller blade on the stairway wall. "Apparently he once trained in the plane and used it during the 1930's before he joined the RAF."

"Who was he?" I asked, trying not to appear too thick, but his name never rang any bells. "Cats Eyes Cunningham. You haven't heard of Cats Eyes Cunningham! He was one of the most celebrated pilots of the war. A hero to us kids. He was famous for being the first pilot to shoot down an enemy aircraft at night. He flew a Bristol Beaufighter and shot down at least 14 planes during the Blitz. When the press asked him how he did it, he told them that he ate lots of carrots to help him see in the dark. They say that is what started the rumour about carrots being good for your eyes. After the war it turned out that he had a little help. In his plane he had one of the first ever aircraft radar systems. He knew where to look as his machine was picking up the enemy planes. Of course he couldn't tell the public so he stuck with the propaganda story of the carrots, and kids have been eating them ever since."

My next customer took me an age to find. She had given me instructions, but trying to locate a small dirt road off the main Dallington Road was proving difficult. Eventually I spotted a small track and pulled into it. I parked and phoned my customer but I had no signal. I looked into the bleak woods and decided that it was worth a try. I drove for over mile down the bumpy woodland track, over small streams and the roughest terrain, praying I was heading the right way. Luckily my old Land Rover gobbled up the track with ease; her new tyres were doing their job. I expected to hear banjos spark up because it looked like I had driven into hillbilly land.

At the end of the track I was delighted to find a small hamlet of houses, possibly some old village that had slowly died as roads and railways bypassed it. I got out of my car and looked at it as I pulled my tools out of the back. I had spent two hours washing and polishing her on Sunday and now she was filthy again. To make things worse I stepped backwards straight into a puddle. "Now your shoes are as dirty as your car," came a voice from behind me. It was my customer smiling from her doorway. "I'll wipe those over for you while you see if you can save my machine. I can tell you, when I broke it I cried for two days. I am so glad you found me – even the postmen get lost here!"

I was expecting some old hand sewing machine but I was wrong. On the table sat a brand new super-expensive computer multi-stitch. She had brought it back from America and just plugged it into the mains. She had not realised that we run on 240volts and America runs on 110volts. In a second she had blown her brand new prize. No wonder she cried so much! Luckily for me the machine was protected and inside was a fuse panel so I soon had the machine running like new. What am I saying, it was new! I left that job with a smiling customer and clean shoes.

And so the calls went on; one customer out, two plastic throwaway machines irrepairable, all the usual stuff that makes up most of my normal working days. Until that was, I met the vampire sausage!

"BILL, GET THAT FILTHY THING OUT OF HERE!" shouted the woman. She was talking about her bog-eyed spaniel. It had all started out innocently. I was fixing the woman's sewing machine when the strangest looking dog I had ever come across rushed up to me, wagging its tail. Imagine a bald sausage with huge eyes and fangs and you would be nearly there. The dog was completely bald save from a few hairs on its tail. "It had a skin complaint," said my customer as she saw me staring. "We have treated it with all sorts of pills and steroids and at last its hair is starting to grow back. Bloody thing has cost us a fortune. I only keep it because of him" she added, nodding in the direction of her husband. I stroked the dog and it seemed to love it so I tickled his ear. Big mistake. The next thing I knew the mutt had taken my ear-tickling as a sign for sex and has started humping my leg like no tomorrow.

The sight was one that would make you pull a face, as if you are sucking lemons while being shocked at the same time. There I am looking down at this creature from some horror B movie having the time of its life. Its boggle eyes are staring at me and it is now dribbling and I swear smiling at me. Its tongue is hanging out one side of its gob as it is panting like a steam train. I am shaking my leg to try and remove it, but I think it is enjoying it even more. Maybe he thinks I am participating? It seems to have a strangle hold on my leg and I can feel its warm, foul body pulsating. The woman is screaming and I am shuddering.

"BILL, GET THAT FILTHY THING OUT OF HERE. Take it for a walk and put its coat on first. The last thing I want is more vet's bills." With that, her husband grabbed the humping sausage and managed, with some difficulty, to disentangle it from my leg. Bill ran off to the hallway and started putting the dog's coat and lead on. I was in a state of disbelief; why are so many spaniels like that? The door closed and peace returned to the home. I sat down, shook my head and carried on with the machine. The woman went to put on the kettle – we both needed a drink. "They only have one brain cell between them those two," she shouted from the other room. "Sometimes the dog has it, and sometimes my husband has it, but never together. They drive me to despair."

As I drove away I spotted Bill, with his humping sausage, wrapped in a warm tartan dog coat. Bill had somehow managed to get dog's muck on his hand while he was clearing up after the mutt. He was rubbing his hand on a tree as I went by and did not see me. The dog however spotted his lover escaping. It looked at me with its huge boggle-eyes and started wagging its tail! I shuddered and accelerated away.

I popped down to the coast and in to see Rose, along Valley Road in Peacehaven. I have to call on her every year or so as she manages to tangle up her old Singer 99k. Valley Road is down a long, rough track and I remember Rose telling me that for two months in winter they get no sun at all. It was like being in Siberia. The valley freezes and all life seems to leave. I knocked on her back door and heard her call out. As I opened the door I sniffed the air. Nothing, damn. Quite often Rose is cooking; she is the Cake Queen of Peacehaven and a slice of her cake would have been a welcome sight. "Alex I am just making Welsh Rarebit. Would you like a couple of slices?" Would I! Call the preacher, we're getting married, I thought.

"Yes please I am starving."

"Well you get on with my machine and I'll have them done in the time that you can whistle *Rule Britannia*." I looked into the creamy mix that Rose was stirring up as she dropped a pinch of cayenne pepper into it. "I ran a café for 25 years and Welsh Rarebit, well my version of it, was my most popular dish."

"Is yours different to normal then, Rose?" I asked.

"Not really. Some people call it 'tom toast' or 'buck rabbit' – it is a bit of both. I mix grated Welsh Cheddar with a dollop or two of Lea & Perrin's Worcester Sauce. I add a splash of cream and a teaspoon of Colman's English mustard and a pinch of cayenne pepper. I spread it onto some toast and add a slice of tomato to one corner and slide it under the grill until it is bubbling brown. I then pop a fried egg on top and you have a five minute meal that people adore."

"Oh Rose, I think I am in love again."

"You promised to marry me last time you had a slice of my Victoria Sponge. I am getting the feeling that you are one of those charlatans," laughed Rose as she spread the mixture over the toast. "Now you get on to my old Singer and I'll bring in lunch."

During the dark winter months, when the forest was bare and little gardening could be done, Rose would spend her time quilting, patching and feeding the pheasants that wandered up to her door. Lunch was amazing and I soon had her machine purring. "I need a reel of thread Rose," I shouted to her. She came running in.

"I have it here somewhere. I put it in my handbag this morning when I was upstairs."

"Rose," I said gently. "You're not carrying a handbag!"

"No silly, I call my bra my handbag." As she spoke, she rummaged around down her blouse pulling out a pen, a packet of radish seeds and then a small reel of white thread. I had to hand it to the old girl, she was still making me giggle. She passed me the thread, wiggled her bra about and went off towards the kitchen.

Before long I was ready to go. "No charge Rose," I said finishing my last dregs of tea.

"I know," she replied with a smile. "The last time you charged me was back in '89. That was 24 years ago. I still have the receipt, I kept it with the machine. I often wonder how you stay in business. Especially with all the perks you leave me!"

"What on earth are you talking about now?"

"Haven't you noticed my lovely body warmer? I have been wearing it every winter for over eight years now. You left it here, over the back of the chair, and never came back for it. It fits me like a glove," she continued doing a twirl in it.

"So you had it! I phoned so many customers but I never tracked it down. I'll tell you what, looking at it now I think you had better keep it. It looks as if it would be a little snug on me."

"You bet I'm keeping it. Builders' rules you know. If you touch something twice it yours! Or mine in this case."

"Well that suits me fine. It must have taken me all of 20 minutes to fix your machine and that cheese on toast was worth at least a million pounds. Look at the size of that." I interrupted myself pointing to the largest pheasant I had ever seen at her back door.

"Oh, that's Phil the Pheasant. He is a greedy so-and-so but isn't he so beautiful."

"I bet he's tasty too," I laughed.

"Be off with you," said Rose waving me out of the back door, "And don't let Phil hear you talk like that."

I headed along the coast to a call at Newhaven Heights. The woman had bought one of those large mobile homes that they cleverly referred to in the advertising brochures as 'a prestigious retirement development'. Her home was no more that 15 feet from the cliff edge and as I got out of my car I could feel the force of the sea pounding at the base of the cliffs below. "Oh they think I'm stupid, but that cliff edge is going to be here long after I'm gone. All I wanted was a nice place to spend my last years and this is it. I can see the sea, feed the gulls and I paid less than half of what they were asking. When I'm ashes they can all squabble for my little plot. Oh, you may have to wash your car. When it's rough like this, the waves hit the chalk and the wind brings chalk spray over the top, so your car will look like a frost has settled on it."

"Oh don't worry about my car, it already has half a forest track over it and an inch of mud. I think if I put it in the sun it would crack and peel." As I left, I could see a white chalk frosting on my car like icing. I had created my very own mud cake car and iced it to boot.

My day was coming to an end and thoughts of a lovely meal were enticing me forward. The sun was setting and throwing amber rays over the South Downs, creating a stunning scene along the Cuckmere Valley. The sheep were casting long shadows and had become the colour of mud as the hard winter had covered their thick coats. Sheep do several funny things; they run and huddle in corners hours before a storm, looking like they are having a last minute team talk before a rugby match. They march in perfect lines following each other from one end of a field to another as if they were being drilled by Sergeant Major Mutton and they always put their bums into a strong wind. It is easy to check which way a fresh wind is blowing, just look at the sheep. As the cold wind was hitting the flock that I was passing, their wool was parting like the pages of a book showing their thick cream wool beneath. I love sheep – whatever happens around them, they just keep on eating as if nothing must get in the way of another mouthful of grass.

I rolled along the top road, past Friston Church perched on the summit of the hill by the pond, which was the colour of dirty dishwater, and dropped down the winding road to East Dean.

Passing 'The Tiger' I looked across at the new Thai restaurant. I had made a reservation there for our anniversary meal later, and within a few hours Yana and I would be sitting at one of the tables. The new restaurant had taken over from a restaurant called Grimaldi's – probably named after Grimaldi the Clown by its half-Italian owner, Diana De Rosso. Inside her restaurant there were several toy clowns and masks. Diana was a large older lady, and she ran the restaurant for many years. She was a very popular sight and a little eccentric, living with a parakeet called Tacie, her cats and an old grey parrot she used to chase all over the place, called Joe (probably after Joseph Grimaldi). She also fed the local wildlife, and everything from badgers to families of foxes turned up at her doorstep.

Grimaldi's was a great place to eat, but locals knew to keep away from Diana's 'pie of the day' if she was helping with the cooking. What most people never knew was that the sweet old dear who loved animals was a spy! Yes, a real certified spy.

Diana was born after a liaison between her mother, who was a beautiful ballroom dancer, and an Italian count who was, by all accounts, an adorable rogue, gambling and womanising across Europe on his way from one horse race or casino to another.

Diana grew up in the glitzy world of thoroughbreds and gambling establishments. By the outbreak of WW2 she was an accomplished opera singer. In 1941 she was approached by British Intelligence. Some say she was already quietly working for French Intelligence as she had spent many years growing up in France and spoke the language fluently. British Intelligence wanted her to move around the neutral countries in Europe. She could then mix in the right circles, picking up intelligence that could be put to good use. To allow her more freedom in Europe, they came up with a plan for her to marry in order to gain precious papers. Her brother-in-law at the time was James Mason, the famous British actor. He knew a Spanish actor from a good family who was desperate to stay in England during the war, and he agreed to the arranged marriage. This allowed her to gain the international papers and travel permits. A temporary wedding was arranged, and before long she was married to a stranger that she only ever met once more during her entire life.

Diana was then put to work gathering information from all over Europe. She would perform in neutral countries and entertain important guests at her after-show parties. She would also attend any important bash that she could, all the time gaining snippets of information. The information would be hidden in plain sight on her music sheets disguised as song notes. No one ever bothered to successfully question the famous singer. Funnily, she would have been in big trouble if they had, as she apparently never learnt to read real music but actually memorised her musical recitals by listening to recordings.

After WW2 she carried on her missions, but this time it was behind the Iron Curtain, sending back anything useful to help with the Cold War. She continued visiting much of the old communist bloc right up until she grew too old to travel. As time went by, Diana would wander around the village in her strange old clothes or sit above her restaurant, where there was a sweet roof garden stretching over the garage, and play with her cats. She appeared on a TV programme in the 1990's telling us a little of her wartime exploits.

I was told that her life came to an end as bizarrely as she had lived. Early in February of 2003 Diana was chasing her African Grey parrot, which she used to let fly free around her house, but for some reason he had refused to get back in his cage. Joe was a bright parrot and was apparently answering her back in French as she chased him! Diana fell and died shortly after. What an amazing woman.

I buzzed up the hill to Eastbourne with just one call left of my day, and then dinner. The last call started like the first. The couple were in full fettle, chatting away as I took her old Bernina apart. Someone had made a beautiful job of fitting a non-standard motor to the machine and making all the brackets and connections. When I quizzed the couple the husband confessed. "Yes, I did that. It has been working for over 20 years so it must be okay."

"Oh yes, you have done a great job," I replied. "Were you an engineer?"

"Sort of, more in the electrical field really. We put together some of the first radar."

"Cup of tea Roy?" Asked the old dear, as she came in to the room with mine and placed it on a small mat on the window sill.

"Yes dear that will fit the ticket." Off she toddled while Roy continued.

"I remember when we had to pick up some Magnetrons. Sounds like something out of a Dan Dare comic but the magnetrons had been flown down especially for us for D-Day. It would allow us to convert our radar to shorter frequency which would be more accurate, especially for detecting the smaller guns and tanks. I didn't think too much of it at first, just a little jolly in the truck, a day out with the lads, but when the armed escorts joined us front and rear, I started to realise something was up. The Magnetron was super secret, beyond top secret. The trip to the air base and back was a bit hairy I can tell you, the roads were checked and junctions patrolled. It wasn't the jolly I had imagined, and by the time we returned I had aged ten years. Still we got them back safely and fitted the units to our equipment and they worked a treat.

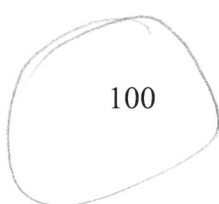

"We went over on D-Day to Sword Beach with the 3rd British Infantry. It was pretty hairy stuff. The tanks went in first with some commandos landing by air behind enemy lines and fighting back to the beach. We were supposed to capture our objectives and link up with the Canadians on Juno Beach as soon as possible, however by the end of the day we still had not managed it. Our planes kept most of the counter attacks at bay as we cleared the beach and set up our radar. Once we were up and running we could work out which planes were which.

Ours carried a special signal so we cleared them from as much ack-ack as we could, same with the ships. We could also talk down friendly planes as we could see exactly where they were heading.

"As night fell we were still stuck on the beach and worried about a counter attack from the sea, so we crept down after dark and set up our two big platforms pointing out to sea. If we pointed them slightly down we could even tell if any U-Boats were stalking the depths, that's how good the Magnetrons were. By the morning we had connected up with the Canadians and started our move inland. The rest, as they say my boy, is history. Now where's my tea."

Roy slowly staggered off to find out what had happened to his cuppa and I was left stunned at the way that he had just relayed such an amazing piece of history, as easy as if he were making a sandwich. Old soldiers, I thought, shaking my head in astonishment as I finished off a row of stitching. You see these old people everywhere, usually jamming up a queue or slowing down the checkout somewhere. It is one of those amazing facts that we owe our way of life to those old fogies who are the hidden heroes of our society, and before long every single one will have gone.

When my daughter Sarah was working as a professional photographer she used to take a lot of pictures of the First World War veteran Henry Allingham who, as he got older, became a local celebrity, then a national hero. At one session Henry was being bombarded by people trying to get their picture taken with him. After a while he waved Sarah over and whispered to her, "Who's that man who keeps pushing in next to me?" Sarah whispered in his ear, "That's the future Prime Minister, Gordon Brown." He wasn't impressed!

My work day was over and the Land Rover almost sighed to a halt as she rolled into her usual parking spot. I had parked in the same place with her for 17 years and the ground was slightly dipped where her wheels sat. I ran inside and chucked my jacket away. "I need a bath," I shouted to Yana as I ran upstairs to put the water on. "You wouldn't believe all the stories today. I met a dog that looked like a skinned pig on sticks out of an x-rated horror movie. Oh, and my trousers need washing. Don't ask what's on them – I'll tell you about it, over dinner."

SUNLIGHT ON A SPECIAL DAY

Through ancient glass the sunlight streams,
It shines today upon a couple's dreams,
And upon the walls of a holy place,
And upon a bride in satin and lace.

The sunlight falls on young and old,
And shines upon two bands of gold,
And upon the path it lights the way,
For two people on this special day.

Through ancient glass the sunlight pours,
And shines upon our hallowed laws,
And upon the bonding of two souls,
And on fulfilment of their goals.

The sunlight gleams on christening water,
And shines upon the sacred alter,
And upon the mothers with tears close by,
And on proud fathers who dare not cry.

But most of all the sun does shine,
Upon the joining for all time,
Of young love, which has now begun,
A journey with two hearts as one.

Scotney Castle is a magical place on the edge of my area. Each time I visit, I feel like writing a fairytale.

Here is a view of the sweeping Eastbourne coastline showing how it curves in and out, one bay after another.

The sea has always fascinated me, and stormy days when the seagulls shriek at the sky are my favourites. Living on the seashore at Pevensey was pure heaven.

ABOVE: Sunrise over the sea along Eastbourne seafront can be majestic, especially with the beautiful lines of Eastbourne Pier.

LEFT: This is a picture of my first paid job as a page boy. I was five years old, and at some point I was messing around and fell into the small pool at the church. Typical me!

LEFT: Book signing for my American Publishers. It sounds glamorous but in reality, you are stuck in a strange place, often on an uncomfortable seat talking to strangers for hours and spelling their names wrong!

RIGHT: Against all global warming predictions, the last few winters have been the coldest I have ever seen in Sussex. I am starting to believe my old teacher at college who told us we were entering another ice age. The tallest person in the picture is Dominic who is six foot seven, so you can see how big the snowman is.

ABOVE: This is a statue of Sir Arthur Conan Doyle in Crowborough High Street. He often gets traffic cones put on his head.

BELOW: Here I am in my garden. As I have aged, I have enjoyed gardening more and more and I am happiest here, just pottering around amongst the birds and the bees.

Although Normans Bay is perched between the bustling costal seaside towns, it is as remote as the North East coast and a photographer's dream.

Sluice Lane, the road to Normans Bay is so remote that rubbish and burnt-out cars often decorate its hedgerows. I went by this one as it was still smouldering.

These are the last remains of the SS Barnhill, sunk by the Germans. Three fishing boats pass its boilers as they leave Sovereign Harbour. I was fishing for bass when they came by.

Here is the SS Barnhill still aflame and in the process of being evacuated by the Eastbourne Lifeboat.

ABOVE: This is the certificate of Henry Blake from the story 'Days of our Lives'. A fascinating story from Sid Day.

LEFT: Here is Sid Day and his wife Eve from the same story, enjoying a stroll along the seafront in their younger days.

ABOVE: *Here we are beach fishing again, this time with Dominic and my old school friend Andy Russell. The sun has just set and within the hour the sky would be lit by the stars and our lamps.*

LEFT: *Here is a picture of Cats Eyes Cunningham from the story 'Coral'.*

Even in black and white, harvest time in Sussex is inspiring. You can see why so many songs and poems are about that time of year. Here we are looking across to Alfriston and Lewes, lost in the folds of the South Downs. My daughter Sarah took this picture late Autumn in 2012.

Here I am, driving along a back road from Ripe towards Arlington. I often bump into all sorts of animals wandering along the narrow lanes early in the morning. There was no farmer in sight – they had just escaped from a field.

THE RAINMAKER

Grass had shrivelled to a dusty brown,
The countryside cloaked in a hazy gown.
Summer had come for far too long,
Warnings had been in the Swallow's song.
But no one wanted this eternal heat,
Sun-baking, burning from head to feet.
Nowhere to hide no place to go,
To find fresh air or cool waters flow.
Villagers gathered, something must be done,
To rid us of this infernal Sun.
Where was the water on which they relied.
Crops had withered, cattle died,
One spoke out "We must swallow our pride,
Call the Rainmaker," they all cried.

Word was sent throughout the land,
With hope, prayer and outstretched hand.
To find the man they needed so,
Who will bring rain to make crops grow.
The families all huddled, some prayed,
A few worked on with tempers frayed.
Would he come this man we need,
To make ground damp for fertile seed.
Many days came and many went,
A thousand fretful hours were spent,
Watching, waiting for him to appear.
One shouted, then two, then a cheer.
The Saviour was coming, he will be here,
The villagers assembled as he drew near.

A giant of a man with dark sunken eyes,
He looked then listened to their cries.
"Just one demand for my deed,
The first born of every breed".
The meeting was short, the agreement fast,
But their memories were poor not made to last.

"Come next spring I shall return,
Make sure they are ready or all will burn.
For if a bargain I do make,
Then no man or beast may it break."
And with those words his head flung back,
Arms stretched upward with a thunderous crack,
Dark clouds rumbled from a clear blue sky,
And rain drops fell from up on high

And down it came, hour after hour,
Controlled by some unearthly power.
They all jumped for joy and celebrated,
Gone was the dreadful heat they had hated.
In it's place was the gift of life,
The end to all their dreadful strife.
A million cracks soaked up the rain,
Villagers danced gone was their pain.
So the deed was done, the pact was made,
But as time marched on memories did fade.
The ground was damp, crops grew tall,
Summer swept through in came fall.
The animals grew fat and heavy to pull.
Store grains were high, the larders full,

The Moons swung by and Spring did come,
And all had forgotten, except the one.
The ground rumbled with his roar,
"Give me the first born as agreed before".
But the villagers were accustomed to their gains,
Why give him anything for his pains.
Another meeting, again short and sweet,
With unanimous voice they decided to cheat.
"Yes, yes, yes" they agreed one and all,
The rain was coming anyway was the call.
So they gave him nothing and told him to go,
"The rain came regardless how could you know."
He bellowed "When will you learn you mortal fools,"
It is I the Rainmaker that makes the rules."

Head bowed down the Rainmaker did turn,
He scowled and hissed, "For this you'll burn."
His eyes grew dark, his voice became a roar,
"Now you shall pay as I promised before".
His arms rose high, the air became still,
The very elements controlled by his will.
Villagers ran, but there was nowhere to hide,
The Earth's crust opened as they cried.
Yet all too late the deed had been done,
And as the smoke settled there was just one.
He surveyed the dead and barren land,
Wiped clean the Earth with a wave of his hand.
Then turned his back and strode away,
Off to the next village for payment day.

BALLS

Louisa Price was just 20 when the hat making firm that she was working at in Luton, suddenly switched to making uniforms for the Second World War. Many of the factories in Luton carried on making hats, but for the armed forces, instead of Civvie Street.

Luton in 1940 was still the world centre of the hat trade and hats from Luton were worn by everybody from cowboys to Cossacks.

The hat making industry in the Luton area may have started as early as 1610 and over the next 400 years, hats became synonymous with the town. By 1680, thousands of people in the Luton area were employed making hats. The pinnacle of the town's hat industry was in the late 19th century.

No one is exactly sure why the hat industry blossomed in Luton, north of London. There are a few fanciful tales and legends, but precious few facts. One legend tells that Mary Queen of Scot's son,

King James, brought the trade with him, when he claimed the crown of England after Queen Elizabeth's demise.

It seems a far-fetched story but there must have been a reason for Luton to be the centre of the hat trade besides its closeness to London and its abundance of raw materials.

What we do know is, when expertise in a field was learnt, and travel was difficult, that knowledge stayed in an area. Much like the cycle industry in Coventry or the needle industry in Redditch. Hatton Garden is still a diamond centre and Brick Lane in London, where some of my French Huguenot family came to in 1685, has been one of the best places to buy silk for over three centuries.

From the 1850's, a process called 'carroting' was used in the making of felt hats. Animal skins were rinsed in a carrot coloured solution of a mercury compound. No one realised the deadly consequences of mercury on the workers.

This 'carroting' process separated the fur from the pelt and bonded the skin. The vapours did not seem too bad. However they were actually highly toxic leading to a lot of mercury poisoning in the hat industry.

The psychological symptoms associated with mercury poisoning led to workers behaving erratically in an almost insane manner, which is where the phrase "As mad as a hatter" comes from. By 1941 the mercury process was banned and mad hatters disappeared into the children's books forever.

The hat trade that is left in Luton today is only a tiny fraction of a once huge industry. Everyone used to wear hats, especially the men. Even in my youth, it was rare to see a bare head. I remember seeing pictures of London in the 1950's where thousands of people were rushing here, there and everywhere, and nearly every head, rich or poor, had a hat on it.

At the outbreak of World War Two, officials came to Louisa's hat factory and made all the workers sit tests. Because Louisa passed the test, which she assumed was some sort of IQ test, she was moved to a factory set up to manufacture ball bearings for military vehicles, possibly a small part of the huge Vauxhall factory that was in Luton.

On the first day of her new job, she lined up with all the other girls and was inspected. Louisa had dark eyes and a high IQ, both essential for one important job, inspecting ball bearings. Dark eyes were supposed to be stronger!

Louisa was then trained in the quality testing of ball bearings. Her teachers constantly impressed on her the importance of ball bearings in the war effort. Once her training was completed, she was given her own testing laboratory which was a small room with the sides set up with various containers of liquids to clean and stain the ball bearings before microscopic examination.

As she started her first morning she kept in mind what her teacher had told her, every machine, every tank, airplane, every car, train and truck ran on ball bearings. They were absolutely vital. If one ball bearing failed, so did the whole machine. If Louisa did not do her job properly, Spitfires would fall from the sky, tanks would grind to a halt, and Louisa would singlehandedly stop the Allied effort! She was petrified and made a promise to herself that she would do her job perfectly.

On her first Monday, she put on her new white coat and thought to herself how important she looked. Her first job of the day was to walk around the factory with a shopping basket and collect random sample bearings from all the machines that were making them, carefully noting where each specimen came from. Then back to her lab and down to business. The ball bearings needed to be de-greased first, so Louisa pulled on her long acid-proof gloves and dipped the balls into the special cleaning fluid. Next was to stain the balls with acid and another chemical and then coat them in something that made them yellow before drying them. Finally, Louisa would get each bearing under the microscope and examine it for any signs of imperfection or manufacturing malfunction. All the details were listed and date stamped.

Louisa was doing well on her first morning and loving her new job, when two young Irish lads turned up with containers for cleaning fluid. "Morning beautiful, we have come for the cleaning fluid."

The laboratory was quiet compared to the hustle and bustle of her old factory, but a while later, two more smiling men turned up for more cleaning fluid, this time with a flask and a pan making some poor excuse about the containers. This is a great job and everyone is so friendly, thought Louisa as she chatted to them and told them how she was doing. This procedure of interruptions carried on at regular intervals throughout the morning, with each visitor taking great care to seal the cleaning solution container up after taking their supplies.

After lunch Louisa noticed that she was running low on cleaning fluid and went to her foreman to ask where the supplies were kept.

"That's not possible girl, you have at least a month's supply! Did you knock the drum over?"

Louisa explained that although she had used some of the fluid, most of it was taken by the factory workers and friendly Irishmen in all sorts of containers, apparently for important work around the plant.

"Heavens above girl. Your cleaning fluid is medical quality alcohol!" He squirted out. "Never let it out of your sight – let alone the room."

"No one told me," Louisa murmured. The foreman grunted something unrepeatable. Louisa and her boss then set off to find out what was being done with her supplies.

As they walked around looking for the culprits they heard laughing coming from inside one of the large storage facilities and went to investigate. They found around a dozen men, drunk as skunks, sipping the fluid and singing songs. They all cheered Louisa as she entered and raised their drinking utensils – an assortment of mugs, cups and jam jars. The foreman shouted at the men and they noisily filed past back to work, some winking at Louisa as they went by.

Needless to say, Louisa learnt her lesson and the next day, as much as they pleaded, she never fell for the same old blarney again.

After the war Louisa switched back to her hat-making firm, as if she had never been away.

What a great story, don't you just love them, real life, real humour, in the midst of adversity. Thank you Louisa.

The last time I visited my old dear she was 92, living in Bexhillon-Sea and still going strong. She had the habit of patting my face when I repaired her machine as if I had been a good little boy. I just smile. Her stories are worth it.

WHERE ARE YOU?

The blossom has all been wasted
The daffodils search in vain,

For spring is lost in mist this year
And endless dreary rain.

Come on Sun where are you!
It's time to show your face,

To sprinkle our world with sunshine
And brighten up this place.

ISABEL

(WELCOME TO THE FLATLANDS)

Hnaef, the young King cried out to his warriors: "This is no dawning to eastward, no dragon flies here, nor do the gables burn upon this hall; but here battle goes forth, carrion birds are singing, the grey-cowled wolf howls, the battle-wood dins, shield answers sword. Now the moon shines, wavering under the clouds; arise to deeds of woe, which will draw down strife upon this people. Awaken my warriors take hold of your shields, resolve upon valour, fight at the forefront, strive in the spirit!"

CHAPTER 1

This is my favourite tale in the book, and the longest, so I have split it into chapters. It is really a mini book by itself. In our story we are going on a journey to a fascinating piece of my world, no bigger than a thumbprint on a map. It has played its part in some of the greatest events in British history and yet today it is spectacularly overlooked. From vast invasions to dark tales of smuggling, through thick and thin, this place has stood against the forces of man and nature. My story winds and wriggles like the roads we are going to travel along. Where are we going to I hear you ask? We are off to Normans Bay. Here, as a young man, I spent a happy few years learning about some of the characters that shaped this tiny coastal village, like Auntie Brim who ran the village shop and the Emery Brothers who cut a living from the unforgiving flatlands.

I often stand on the shoreline at Normans Bay gazing out to sea – it is one of my special fishing spots. I watch my rods pointing at the horizon and let the waves of history flow over me. The sea, unlike the land, hides its scars. The sea lives in the present, its history lost in a moment, but over these waves the invaders came, the grasping Romans who conquered, and took wheat, lead, gold and much more as they pillaged the soil of Britannia. They smelted iron to power their endless war machine and carried it to Rome on the backs of slaves captured from our land: they plundered Britain long before the Saxon and Celt.

The Vikings came with the rolling mist in their longboats, their oars arching to the rhythmic beat of their boar-skin drums. They also wanted our ancient land. At Pevensey there was a great slaughter during the dark ages, hence the poem at the start of our story.

Then the land-grabbing Normans arrived with promises to God coating the edges of their blades. The Spanish Armada, Dunkirk and much, much more – it's all here on the shingle banks of Normans Bay, the past that slips almost unnoticed through the pages of our history.

The name Normans Bay, originally named 'the-Bay-of-the-Normans' is self-explanatory. The shingle bank is where the Duke of Normandy landed most of his troops in 1066. Many say it was at Pevensey Castle where he landed but there was once a huge inlet here leading up to a large river that ran most of the way up to the market town of Hailsham. He would have landed his ships all along the inlet and the bay. Legend tells that, after landing, he dispatched many of his ships along the coast and burnt some in front of his men to let them know there was no going back; it would be victory or death. William was not going to let anything stand in the way of what he believed was his country, his promised land. Even when he tripped getting off his ship, stumbling into the salt surf he claimed it was an omen of good fortune, exclaiming on his knees that this was his land, shouting to his men that they had landed on this soil with a mission from God.

He left just a handful of ships with supplies at Normans Bay, fortified the old Roman castle at Pevensey, and then moved along to Hastings, stopping briefly to decimate, amongst others, the ancient village of Crowhurst. After his famous victory on the ridge at Battle, known as The Battle of Hastings, it was onwards to London and into the history books. More of William later.

CHAPTER 2

One night early in February of 2012, a mini miracle happened and nowhere looked more spectacular than the flatlands around Normans Bay. The night was so cold and clear that Orion the Hunter hung in the stars close enough for me to reach up and pluck him out of the night sky. An unusual combination of weather systems had turned all the moisture in the air to ice crystals; it was a rare and spectacular hoar frost that I had only seen once before in my life. The next day rose bright and clear and every bare branch, bush and tree was coated in frozen particles. Without a single snowflake falling our world had turned diamond white and East Sussex had walked straight into the mythical pages of Narnia.

The global warming experts were once again dumbfounded and scuttled away to modify their explanations and wait for a few warm days before scaring us to death again.

March then came in with a bang. The previous November had been the wettest since records began, December the coldest for over a century. January had slowly clawed its way into a bitter February and summer seemed to be just a dream from ages past.

Our hard long winter brought with it a beautiful parting gift. The most splendid spring in living memory arrived. All the flowers that should have blossomed early were late due to the frozen earth, then as the weather turned almost instantly to a balmy spring, all the later trees and flowers, triggered by the sunny skies and warm breezes, blossomed early. For the first time in my life the whole countryside was ablaze with colour as three months of blossom all arrived in the same fortnight.

That superb spring the sun had called to the celandine, and as sun lovers, they obligingly opened their yellow petals in worship. Daisies sprinkled the verges in numbers too huge to even guess at. The early Saxons called them the 'days-eye' as the pretty little weed resembled the rising sun or the eye-ofthe-day. A modern invader was also busy, the Danish scurvy grass – the pretty verge-hugging weed had painted many roadsides with its colour. It is normally a salt-marsh plant but in my lifetime it seems to be moving inland at some speed. Experts say that it is the heavy winter road gritting with salt that had allowed the low growing, pretty pink and white flower to prosper. Whatever the reason I am pleased that it has decided to move here and become so prolific along the roadsides, where it now flourishes in an environment that few other plants can survive.

Fields of rapeseed, the modern bio-fuel miracle, had turned the countryside yellow, as had the dandelions which, spared the executioners blade of the well-kept town verges, had flourished beyond description. The French named the sharp little petals of the dandelion 'Dent-de-Lion' or teeth-of-the-lion and, though a little altered, we kept its name. As a child I remember well the gypsies picking them for their medicinal brews and salads. Any youngster knows that when the dandelion turns to seed it becomes a clock – it can be surprisingly accurate, especially if you know the time before blowing at the little upturned umbrellas. Within a few days the countless yellow heads of the dandelion turn to seed and the fields, almost overnight, become white as if covered in a low mist. Then, as the lollipop heads dance in the wind their seeds fly, leaving fields of spikes as the only witness to the amazing colour that was there a few days before.

For many months of the year the colour of England is drab grey, battleship grey. The sky and sea often match so seamlessly that sometimes it looks as if the ships are floating in the air, so when these precious few brighter days of spring arrive, as if by some miracle, it puts many of us in high spirits. On the first warm day of spring I have seen people that do not smile for most of the year, beaming from ear to ear. Postmen start wearing their shorts and whistle as they work.

And all this happens so fast – one week the woods are bare, the next the branches shimmer with fresh green tinges, hinting of better things to come. One moment the temperature hardly climbs above freezing and days later it is balmy with a warm wind from the South. Anticipation of those first warm days of spring are just delightful. 'April's showers brings May's flowers'. Also there is a power in the spring that the ancient religions, like the Druids, knew all about. Other faiths turn to rebirth but the power is in what some call Gaia, the earth itself. Spring recharges damaged souls and re-balances life.

It is the only time of the year that the countryside appears with the same colours of a fresh lemon meringue pie; the yellow of the celandine, dandelions, gorse and rape seed flowers are topped with the hawthorn and blackthorn blossom which provides the meringue topping. Every bird in spring has a song in its throat and every tree a blossom. Skylarks sing as if they were to live just for the day and seemed never to bother to stop for breath. They sing continuously, perched on the breeze held up by invisible sky-hooks. Never can such a bird have existed that looked so plain but sings so sweet as the Sussex skylark. There is no smell on earth as sweet as early spring, when the land wakes from its long winter slumber, when the soil, warmed by the sun, breathes and the smell of new life fills our lungs.

CHAPTER 3

Our journey to Normans Bay starts at Sluice Road, which leads by Normans Bay and then carries on towards Bexhill. I was having a break before moving on to my next customer, Isabel, at Bexhill, (she will be at the end of our story). At certain times of the day the road becomes a rat-run, with the occasional mirror, wheel trim or piece of bumper plastic to tell of the motorists' close encounters along the narrow road. However nearing midday the road often falls into a blissful silence. On the day I was there, the cattle were laying in the soft grass and a heron stalked its prey in a narrow gully along the edge of a field. There were not many birds, a few crows and circling seagulls riding the high thermals. I clearly remember the thousands of lapwings that used to darken the sky here many years ago. They would rise and fall in great waves before dropping to the fields to rest for the night. Now they are a rare visitor to these famous marshlands.

A few small birds were gathering nesting material to make their homes in the safety of the prickly hedgerows. Sparrow hawks patrol this territory and the spiky hawthorn and blackthorn make the perfect nesting places safe from harm. The ancients of this land called the hawthorn, 'Huath' and it was believed to possess magical powers. Only recently has modern science rediscovered some of the amazing qualities of the hawthorn, which really does sooth the heart. Many an old dear who knew the secrets of the land would mix potions of all the natural elements until their persecution for being witches began. So in tune to the seasons is the hawthorn that it was named the mayflower, and women danced around it and later the maypole to honour the coming of spring. The mayflower is still a symbol of renewal and was even used by the Pilgrim Fathers as the name of their famous ship. The early settlers to Normans Bay would have used the thorns to make hooks for fishing; even today you can easily cut out a thorn, wrap it in honeysuckle twine and place a worm on it and catch your supper. I know, I've tried it.

Sluice Road, is a narrow and winding road bordered by a small copse of willows, high marsh reeds and flat grasslands. To the north lies Pevensey Levels. It is open marshland with many rare plants and insects, like the Fen Raft Spider. It is so unique that it has become a site of special scientific interest or SSI. Mind you, access is so limited you are unlikely to ever see one of the spiders! Joy riders occasionally use the silent road to burn their stolen cars and fly tippers dump their rubbish. However, these flatlands are a little spot of paradise. Clear views in every direction call to you. In the distance the sea meets Beachy Head, rising like a frozen wave and continuing for miles down the coast.

The Normans Bay sign should really read, 'Welcome to the Flatlands'. There are many high spots of land that stand out from the flatlands. These high spots were once tiny islands or 'eyes', which lifted above the huge tidal bay that reached inland. Some were cut off at high tide and even today you can spot these ancient islands by their names ending in 'ey', like Rickney, Northeye, Langney, Pevensey, Montney, Manxey, Glyndley, and Horseye.

Horseye was said to be hardly the height of a horse's head above the seabed on the highest tide. It has a farm perched almost on top like a cherry on a bun and has one of the most perfect views of the Pevensey Flatlands. All along the tiny road from Rickney to Herstmonceux run small rivers and forgotten raised paths that lead to the busy market town of Hailsham, where livestock is still sold on market days.

I was actually parked in Sluice Road recovering from my previous call that had gone wrong right from the start.

I had reversed up the narrow drive to her house and stopped because a long, single five-bar wooden farm gate blocked my path. I got out of the Land Rover and saw that it was an automatic gate with a red button on a wooden post. I pressed the button and hey-presto, slowly the gate magically opened.

That's when the trouble started. I quickly realised that the gate was too long and that it was going to hit the back on my car. I grabbed the gate and held onto it like my life depended on it. I was sliding on the gravel but I managed to temporarily hold the gate back, I could hear the whining of the electric motor struggling. My brain was racing, there was no space between the gate and the back of the Land Rover to allow me to get around to the driver's side. I put one foot on the back of the car to brace myself against the gate, which was slowly winning its tug-of-war. Images of the gate ripping off my spare wheel and bumper were racing through my mind. I had the brainwave of letting go of the gate, leap-frogging over it, and rushing to the handbrake to let the car roll out of the way. I figured I had a few seconds before all hell broke loose. I took a deep breath and went for it. In one smooth movement I let go of the gate, twisted and leapt onto it. Now, when I was 20 it would have been a breeze, but at my age, with my belly, I got almost halfway over the moving gate and no further; it suddenly seemed much higher and I found myself hanging onto it rather than leaping like a gazelle over it. The gate then hit the back of my bumper and started to scrape along it. I was still trying to get over the top when the gate suddenly stopped! I breathed a deep sigh of relief and let my head drop onto the wood where I was balanced like a large pig on the barbecue. I figured I must have blown a fuse or something but then came a voice. "Alex you do give us a laugh," said a woman standing by the red button. "We have been watching you from the dining room window. My husband is still laughing so hard that he can't stand up. The dogs have gone mad and are now running around the back garden like crazy things."

I climbed down from the gate, red-faced and brushed my jumper off. "How did you stop the gate?" I asked. "Well Alex, I pressed the button – the same one that you pressed to open it. One press opens the gate, next press stops it, and the next one closes it: quite simple really. It is lucky that I know how good you are with sewing machines otherwise I might be tempted to ask you to leave after seeing your amazing ability with automatic gates! What on earth made you think that you could leap over it anyway? Now you had better come in. I need to get the dogs back and make sure my husband has not died laughing."

I spent as little time in the house as possible. Every now and then her husband walked past and I could hear him chuckling. No sooner had I left my customer, (fixing her machine beautifully I might add), than I decided to take my break and recover my thoughts on the ridiculous episode. And that is how I ended up in Sluice Road.

It was while I was eating an apple and leaning against my Land Rover, under the small copse of willows that something quite unusual happened. The silence of the spring day was lightly stroked by the buzz of the insects, birdsong, babbling water and just the slightest hum from the distant traffic on the Barnhorn Road. The scene was already close to perfect, something that Mother Nature effortlessly achieves, when a ripple of wind, travelling in the folds of the warm breeze, shook the willow trees above me. They were full of fluffy yellow seed fronds, sticking out from the branches like chubby chipolatas. Clouds of the seeds were sent flying into the air and, almost weightless, they hung like snowflakes, drifting but hardly falling. The tarmac had already almost disappeared under a blanket of hawthorn blossom and the drifting willow seeds added the ideal finish straight out of a Hollywood blockbuster.

A big bumblebee dug itself out of its winter hiding place in the ground and buzzed way up high towards the sun. I was entranced, standing in the middle of my own snow globe and soaking up the exquisite panorama when the faint hollow sound of a cuckoo floated over the meadow. It was followed by the drum of a woodpecker like the fire of an automatic rifle far away in the woods. Woodpeckers find the perfect hard and hollow wood to bash their brains out at this time of year. They are signalling for a mate and it works, their noisy 'peckering' travels for miles. I looked around and behind me there was a large brown cow lying in the field casually chewing grass. She called across the fields with a long 'mooo' as if to say "yeah, great isn't it?"

I am always amazed that such beauty comes and goes as if in tune with some eternal clock. Occasionally in those briefest of moments we entwine with nature and the seconds of our lives, which normally drip relentlessly away, freeze in time. In that special moment we can touch our own souls.

Only one property, an old farm perched on a large plateau of grass-covered rock, breaks the entire journey to Normans Bay. Rockhouse Farm is set on top Rockhouse Bank, an isolated spot with only its lonely views to keep it company.

Beyond the farm, a low Sussex flint wall in poor repair cuts the sea wind. The pancake-flat fields are strung with barbed wire and tufts of sheep's wool snagged in various places. In the summer the fields smell of wild camomile. Drainage ditches and streams that channel the water out to the weirs and sluices moat each field. They control the water levels of the sub-sea fields. In the hazy distance the domes of the Herstmonceux Observatory telescopes lay amongst the woods like great Roman helmets. Once they were at the very pinnacle of cutting edge technology – now they sit like lost giants, waiting for visitors to turn up on sunny days and run around their forgotten hulks.

To the side of them is Wartling, where the hill was hollowed out and used as a communications and observation post throughout the Second World War. I was told as a child that Wartling was one of the most important communication posts on the South Coast. The only sign of the hills special assignment today is a small entrance, but few know the secrets hidden on the many levels underneath.

To the side of Wartling is Hooe Church, where plague once wiped out the entire village. All the wood and straw properties were burnt and the village moved a mile away where it still is today. The stone church stayed, now isolated, as the sole reminder of those dark days.

Swinging around, like a camera obscura from my panoramic position, you can see the shoreline and Coast Road running along with the uniform peaks of the rooftops silhouetted against the sea like a Monopoly set.

Half way along Coast Road was my old home, where I would fall asleep to the sound of the lapping waves caressing my brow. In the early morning the cries of the gulls, carried on the folds of the wind, would creep into my dreams. My bedroom was so small that there was not enough room for furniture or a table, so next to the single bed was a folding shelf screwed to the wall where I could put a cup of tea. That tiny bedroom was all mine and I slept there like a baby. Although the bungalow, perched on the artificial shingle bank was compact, Dad had the largest garage in the road with the first up-and-over electric garage door in Pevensey Bay.

When his baby business boomed he bought a new Jaguar but the salty sea mist rusted it. Within months it was replaced with a new one by Caffyns, the Jaguar dealers in Eastbourne. From then on, as most of the locals had no idea how to pronounce my Dad's Russian surname, he was known locally as 'Two-Jags' (long before John Prescott took his title).

Living on the seashore was as close to heaven as I will ever get. The sea was my larder and the countryside my playground. That sounds just so gushing, but it was true. The sea provided a huge bonanza of delicious food. Whelks and winkles, washed in on the tide after the storms would cling to the rocks. At low tide the bay was alive with free meals. Cod in winter, bass in spring, mackerel and sole in summer and countless other fish. There were lobsters and prawns in the rock pools and brown shrimp all along the shore in less than a foot of water.

On the Eastbourne Road, just as you leave Pevensey Bay, was a clever old man known as 'Light Fingered Fred'; not because he nicked stuff but because he was so nimble with his stubby fingers and a holly needle. He made nets – prawn, shrimp, lobster, you name it and he would make it for you. The wiry-haired old sea dog sat outside his house watching the traffic pass, stitching up his nets. "Pop down to Fingers, he'll make you a new net in a couple of days for six shillings and a bottle of beer." I bought my first prawn nets from him as well as my first six-foot shrimp net that I used for ploughing up and down the seashore at low tide, gathering my pint of shrimps for supper.

My Dad once told me that the asbestos houses built along Coast Road, which runs almost parallel to Sluice Road but follows the shoreline, were the dream of a First World War pilot who often flew over that part of the coastline on his way to war. He used to look down at the last bit of England as his biplane left the safety of home and marvel at the untouched, sweeping line of the bay. He swore that if he survived the war he would build a row of summerhouses along the coast for holidaymakers, and so he did. They were only meant to be temporary, but nearly a century later, many are still there, asbestos and all.

CHAPTER 4

Back along Sluice Road, The Star of Bethlehem, now simply the Star Inn is, besides the few temporary homes, the only other brick building in the road. In its isolated position it is one of the only pubs for miles around. The white rendered building is low to the ground for protection against the fierce winds that rage for many months of the year. The Star Inn has been perched on the shingle bank for over 600 years. It was originally a sluice house controlling the flow of water from the low-lying fields, which were drained so that sheep could forage.

Wool was the money-spinner of the medieval age and made many landowners rich. Draining the marshland was expensive but it would pay back big dividends. The cash crop was fed on this fertile, reclaimed land.

Water, always being of doubtful quality, was often replaced with the safer brew of ale or beer. Travellers, seafarers and shepherds would pop in to the Sluice House for a pint and a bite to eat. During the English Civil War the house brewed and served its own special beer and started serving it to the soldiers billeted around the area. From that point on it became the Star Inn, with the name that was a nod to the biblical shepherds guarding their flocks.

The inn would pass through time untouched by scandal or history. Smugglers would slip ashore on moonlit nights, rolling barrels of booze up the pebbled beach to the Star. More booty was loaded onto packhorses and ferried inland along back lanes and hidden paths. In 1828 smugglers from Little Common who had landed on the Normans Bay shore to deliver booze, were pounced on by waiting excise men. A fierce running battle ensued into the night ending just north of Bexhill outside the New Inn (which wasn't very new, being founded in 1376). It became known as the Battle of Sidley Green, and besides a violent conflict near Pevensey Bay a few years later, it was one of the last major conflicts between the excise men and Sussex smugglers. Some of the captured men were hung, some of the local smugglers were transported to Australia where they started new lives in the colonies.

Smuggling was, and still is, rife. Billions of untaxed and illegal goods still pour into our tiny island. Today it is mainly drugs but a fair amount of baccy and booze makes its way past the customs – undeclared.

Smugglers are canny lads who know the lay of the land and can move through it silently on the darkest of nights. Normans Bay, like thousands of remote coastal sites, was a prime area for the smugglers of old. Contraband made men rich, and coastguard cottages were often built in isolated spots to allow the excise men to keep watch during the long dark hours, when the smugglers worked best with barrels and bags full of contraband strapped to horses.

Unofficial records on the 'black economy' in 1783 showed that at least one quarter of all the horses in England were being used for smuggling.

CHAPTER 5

Now back to William. I told you he would come up again. Little did the smugglers realise at Sidley that they had fought their pitched running battles just a few miles south from the very spot where William the Duke of Normandy had buried his fallen men after the Battle of Hastings. Harold's men were left to rot where they fell along the ridge of blood (except for Harold himself). A rough compound was built around where Harold's men were slaughtered on Senlac Hill, later to become the spot for the foundations for Battle Abbey.

Harold was another matter. William was well aware of the fondness of the church for martyrs and saints. Edward the Confessor, who had died in the January of 1066 causing all the trouble in the first place, was already on his way to becoming canonised, and sainthood would follow. Abbeys and cathedrals all over Europe held parts of dead saints. Three years after the battle in 1069 a single mummified finger of Saint Germain was taken to Selby to found Selby Abbey in Yorkshire. The year before in 1068, Queen Matilda, William's wife, gave birth to his fourth son near Selby. That son was Henry, later to become King Henry I, the Lion of Justice,

After the battle William had seen his soldiers cutting King Harold up and stealing parts of him to sell. In a rage William swore that any man holding part of the fallen King would suffer a foul fate. A knight called Ivo who was seen hacking at Harold's beheaded corpse was publically exiled in front of his men.

Harold's body was already butchered, but the parts were taken back to where the ships were moored at Normans Bay. The men had special instructions to wait for the lowest tide at night, then when the time was right, they dowsed their torches in the salt water and made their way out to the edge of the retreating sea. There they dug in silence, guards watching for any sign of their presence being discovered. They buried Harold deep in the wet featureless sand, never to be seen again.

Earlier, Harold's beautiful lover, Edith Swan-neck, and Harold's mother pleaded with William for the King's body but William was cold, even when offered the dead King's weight in precious goods he would not budge. There would be no body to rally around, no martyr to worship, and no sign that the usurper king had ever existed. His bones are probably there to this day, hidden somewhere beneath the miles of open beach.

William's fallen men were taken to a spot near Blackhorse Hill on the Hastings Road. Here he knelt for one final time, in prayer for his departed comrades who had helped him win a country. His path was now for London, Westminster and the Crown. They say you can still see a mound near the water tower at Blackhorse Hill where his men lay in silent testament to that brutal day in 1066. A compound was also built around their bodies with guards placed there for four seasons.

CHAPTER 6

Let us continue our journey to Normans Bay. When you look at the little hamlet of Normans Bay it looks fairly modern, but that is not the case. What you see today is far from the bustling place that it once was just a few decades ago.

Originally Normans Bay was almost an island and remains have been discovered dating from the Bronze Age when humans lived close to the shingle beach that dissects the low sweeping bay between Eastbourne and Bexhill.

The place is now like a set from an old Hitchcock thriller filmed in the 1940's, hardly touched by the passing of time and a photographer's dream. Iconic images are everywhere: old broken boats, dilapidated houses, empty buildings, flag masts and nets strewn over the sea kale and samphire. Legend tells that many of the ancient houses that once perched on the shingle bank were built from shipwreck wood. The old café sign can still be faintly seen on the side of one of the buildings, just above the broken windows and flaking paint. The building is a ghost waiting for old customers that never come.

Caravan parks dot the roads leading to the bay and I often wonder what the visitors must think when they find themselves at this remote spot. In the summer they flock to the parks, which are fed by small site shops and the whole place comes alive. The heart of the village spreads landward from one of the imposing circular defensive Martello towers that were strung along the coast to protect Britain from the French during Napoleon's rampage across Europe. A row of terraced coastguard cottages runs along the centre of the forgotten village, which were built on the seashore to keep an eye out for ships at night. Now however, with the constant building up of the shingle from the long-shore drift, their priceless view has been stolen by an assortment of houses of all shapes and sizes, built in front of them on the new seashore. Maybe in another hundred years or so, another row of houses will do exactly the same.

If you stand on the seashore at Normans Bay you will see a small row of beach huts, some tied down with wire, perched precariously on the shoreline. Next to them are the overturned fishing boats and all their fishing debris – lobster pots, nets, crab pots, plastic containers for the fish, floats, ropes and more. A pirate flag often flies from one of the small wooden buildings, signifying the spirit of the remote community of survivors that have somehow managed to escape modern life. Normans Bay is home to an independent and proud hamlet, surviving between the bustling seaside towns along the coast whose chalk hills drop to the sea, each on either side like great arms on some invisible armchair.

The beach is often scarred with the tracks of the huge gravel moving caterpillar machines that constantly push the shingle up to form the sea wall, protecting the lowlands from wild sea surges and winter gales. Worn wooden groynes poke seaward and thousands of thin wooden posts stick up like black sea grass along the shoreline to halt the shingle from being dragged out to sea by the ever-clawing tide.

Normally the only noise at Normans Bay is nature, the sea, the running water from the weir to the outlet pipe and birdsong. It all mingles together in a hypnotic rhythm that soothes the heart. Here the river, Waller's Haven, finally reaches the sea. Into this small river hundreds of fields drain via ditches, streams and channels. Years ago the water used to be helped along the different field levels by wind pumps, which looked like mini-windmills on stilts. When I was a kid, there was just one semi-derelict wind pump still on the marshes, but they are all gone now.

The water collects and runs down into a deep weir on the edge of the village where sluice gates allow it to be set free into the English Channel, but does not allow any salt water back up. Waller's Haven and its surrounding fields were one of my childhood special places where kingfishers fed and nested. The 'king of fishers' still nests in this little haven of rivers where hawthorn and blackthorn bushes provide the perfect feeding perches over the water. Hares used to play in the fields when I fished in the different streams and ditches. On lazy days off from school I would while away endless hours there, in search of the once common eel; the mysterious, slippery creatures from the Sargasso Sea that had drifted across the world, as tiny elvers years later they would be on their momentous journey back to the Sargasso to breed – once they were safely past my worm stuffed hook.

When I lived along Coast Road, (the only other road to Normans Bay besides Sluice Road), I would often fish by the large Normans Bay outlet at night. It was a good spot and attracted plenty of fish. Occasionally as the sun set over the sea it would turn the whole bay into a blaze of gold, as if the sea had been set on fire and then as the sun dipped behind the distant chalk cliffs, the sea transformed into a midnight blue. Then the moon would rise and the sea turned once more, this time to silver. The moonbeams would shimmer on the waves and in my childlike youth I would grab handfuls of them, only to find as I opened my clenched fingers that they had disappeared. Here was heaven, the real heaven, not the one falling from some preacher's page. Sitting alone on the beach under the stars with the lights of Cooden and Eastbourne twinkling in the distance, I would feel as if that hand of God must have passed this way.

When fishing I would sometimes freeze and the hairs on the back of my neck would stand on end. On the wind I would hear songs, like old sea shanties sung by long lost sailors. *"What shall we do with the drunken sailor, what shall we do with the drunken sailor, what shall we do with the drunken sailor eeearlie in the morning."* They would drift in and out with the waves and in my wild imagination they were coming from the drowned sailors and sunken vessels that litter the seabed. *"Fifteen men on a dead man's chest, yo ho, ho and a bottle of rum."* In reality they were probably coming from one of the old fisherman's houses along the shoreline. An expert once recorded over 4,000 shipwrecks along the coast between Rye and Portsmouth, and there are probably many more.

Fishermen still cast their nets here and pull in a hard living from the sea, their tools of the trade clutter the beach. The coastal railway line from Eastbourne to Hastings almost touches Normans Bay and when a whale was washed up in the 1950's, the demand from tourists was so great I was told that the railway put in an extra 'unofficial' stop at Normans Bay and never closed it. This had huge benefits to the small community and the village flourished. Today, mainly two-carriage trains still follow the steel tracks from Eastbourne to Hastings and back.

CHAPTER 7

As with many people, there have been several times in my life when the Grim Reaper has been called away from his normal duties to see if he had an unexpected customer. One of those times for me was here at Normans Bay.

My downfall had started slowly and innocently, but soon changed into a life and death situation where my survival was pure luck. I had a sandline laid out at the low tide mark with 60 hooks on it and I needed to dig more lugworm to bait it up for the following tide. It was a sunny autumn afternoon as I said goodbye to Dad and cycled along Coast Road to the outlet at Normans Bay with my lug-spade tied to my back, and a white plastic bait-bucket swinging from my handlebars. Inside the bucket was my brass Tilley lamp, which I would light as it started to get dark. Dad had mentioned that he might walk along the beach later and meet up with me if he had time. The brief two years that we lived alone together after I left St Bede's, and started training as an engineer at the College of Knowledge in St Anne's Road, Eastbourne, were some of the happiest in my life. It all came crashing to an end with Dad's unreasonable need to live in Spain, but for that moment the long sunny days of youth were shining upon me.

The autumn tides were at their longest and the biggest lugworm lived out past the mussel beds and sandbanks. I dropped my bicycle on the shingle and crunched my way down the gravel, stepping over the rows of hazel twigs that held some of the shingle back. They were a pain at high tide, snagging your line as you wound your fishing line in and I had lost more than one fish and plenty of tackle to the short black beach-daggers. I walked out towards the mussel beds over the first sandbank, rolling up my waders as I 'wooshed' through the sea water in the low lagoons between the banks of sand. On the sandbanks the gentle lapping of the receding tide had made ribs in the sand where the gulls darted about looking for food. Low tide is a wild place; most traces of humans are swept away twice a day and what is left is an ancient scene that our distant Stone Age cousins would instantly recognise.

It was late afternoon and the sun was lazily dropping towards the orange horizon. Gulls picked up mussels and flapped up high in the sky before dropping them on the stones then following them down, dropping like parachutes so that they did not lose them in the maze of weed and rock pools. Young herring gulls tried to copy their elders but many did not follow the dropping mussel closely enough and spent ages searching the debris along the shore where they thought their dinner had fallen. Eventually these clever birds do learn.

There was just the slightest breeze, which was filled with sea scents. Far out towards the horizon a low mist was drifting across the fishing boats, straight out of a Turner landscape. Sandpipers darted in and out of the shallow rock pools turning up weed and grabbing sea lice. I walked over some low rocks where limpets and mussels gathered and through pools, passing a few bits of wreckage left by us just to let Mother Nature know that humans are around. There was an old rope lost from a fishing boat, a broken lobster pot washed up in a storm, a bit of rusty bent metal jutting out of the sand, probably from a boat sunk long ago. Beach combers love this stuff and I can see why, it is a fascinating world down on the seashore and the huge autumn tides show new secrets hidden for most of the year.

I kept walking towards the receding tide, up one sandbar and down through another shallow gully to the next bank. Some of the sand was like muddy quicksand and my waders squished through the clean top layer of sand down into the sticky black mud below. I was used to this and pushed on through towards the clean sandbars, sucking my waders back up with each footstep. About 300 yards out I saw 'Mad Maddy', who's actual name was Madeline. She was beachcombing over some rough ground – if ever there was a 'sea witch' it was she. Her long grey hair was as twisted as coiled springs from the damp sea air and when the wind blew it flew about her head like a furious banshee. She only wore her teeth on Sundays and when she went to the doctor's. It always made me laugh when she wore them, as I could hardly understand a word she said. The mad old dear would throw stones at the seagulls and chase them around the beach, cackling as she went. Dogs would howl at her strange shape as she shuffled along like a crab, searching, always searching. She regularly scoured the seashore with an old hessian bag tucked under her arm, picking up bits of debris as she went. She would take ages examining each piece and then place it in her bag as if she had discovered the crown jewels. I loved the mad old woman and would often sit with her as she chatted about times long past and her lost lover killed in some war or another. She would paint watercolours of scenes on the beach, and if she ever painted a person such as a fisherman, she would always paint them bent over like her. On the windowsill of her beach bungalow were rows of clay pots that she had meticulously stuck all her findings onto with thick, white paste. The pots looked a picture and shone when the sun caught the shells on them.

'Mad Maddy' had two wrinkles on her face for every year of her life, and a gallon drum of Botox would have made little difference to her salt-creased features. I called to her as I got closer, "Careful of the tide Maddy, it's a long'un today." As I approached her she looked up at me with her sunken eyes, hidden deep in the wrinkled pits of her face. She examined me with a gleaming eye, her head tilted like a crow. "You be careful yourself," she cackled, as she walked away barefoot. "I have some cake if you want some later," she added. Maddy did not wait for a reply, she was once more oblivious to everything except the ground that she was staring at and her old bag of treasure that she clutched to her tiny, withered frame. She did make a great fruit loaf but I had little time that day.

As I reached the water line I waved to Ken, who was pushing his shrimp net along in the shallows. We were just too far away to shout to each other so he waved back, holding up a large bag of brown shrimp to show me that he had done well.

I got to the water's edge and stuck my razor sharp spade into the sand, putting down my bucket to roll up my sleeves. I had spent hours cutting down my lug-spade until it was little more than the size of a hand trowel on a long shaft, then I sharpened it to perfection. Cutting through the sand was hard work and the spade had to slice through it thousands of times. After each dig I would re-sharpen my tool like a warrior would his sword.

I surveyed the beautiful scene before me. The shoreline had been made into endless sandy ripples by the receding tide and the sea was a shining sheet of emerald green. I picked up my spade, swung it over my shoulders and did a few loosening swings for my back. I then dropped to my knees and started to dig.

Black lug live in a u-bend of sand and at low tide they drop to the bottom of it to hide. As the water starts to return they move up towards the surface ready to feed. The closer you dig to the sea-line of the incoming tide, the easier they are to catch. Well, easy is the wrong word really; they are hard work and move fast as soon as they feel the first lunge of the spade. I always dig, following the hole down, one spade cut to the left of the hole and one to the right. I needed to move fast and soon I caught a glimpse of a yellow tail disappearing down the sandy hole. I kept at it and quickly dropped my spade, launching my right hand down after my game. I could feel it in the gritty sand. Small shells were sticking under my nails and I pushed further down. By then I was leaning into the hole and the sand was just inches from my face. Then I had it. If I pulled too hard the lugworm will snap in two and be wasted. I teased it out, slowly pulling, and then it came up, sucking out of the hole. Long, yellow and black, my first lug of the day. I threw it onto the hard sand and it stiffened. In one movement I rolled the back of my hand over it, squeezing out its guts. It exploded from its head like a huge teenage zit – disgusting but necessary, as the worms rot quickly if not gutted. I chucked it into the bucket and looked for the next cast. Lugworms are just about at the bottom of the food chain, eaten by every sea creature and they were essential to catch my fish.

I needed 60 lug, but with a good dig on a long tide, I could double it, which gave me two tides of bait and lots of fish for me and Dad. The lug came in thick and fast. That far out on the long autumn tide they were not used to being hunted. The water rarely cleared the sandbank that I was digging on. I looked for nice circular stamps of sand like small walnut whirls that indicated the entrance of the lug-hole. The lugworm that do not go straight down, leave irregular patterns in their 'walnut whirls' telling me that they are going around some obstacle below like a stone or large shell. I always left these as they are harder to dig.

Time slipped by. Every now and again I stopped to survey the scene. A woman was walking two dogs, one was chasing a seagull the other was digging in the sand. Miles of open shoreline lays before my eyes, more reminiscent of the North East rather than our busy coast. The sun had set behind the hills and the sky was losing its light. My eyes grew accustomed to the fading glow and I could still see clearly. I decided to stop for a break and pulled a Kit-Kat out of my back pocket.

After, I opened up the small container of methylated spirits, where a metal clip holds a lump of wick. I carefully positioned it around the central tube of the Tilley light and struck a match. It went out straight away and I struck another, shielding it with the palm of my hand. I touched the wick with the match and a blue flame leapt up the tube, heating the silk filament above. It is tricky lighting a Tilley lamp – it is almost an art. One wrong move and you are covered in meths and paraffin from both the lighting wick and from the lamp. Both stink. I waited till the wick had nearly burnt out and twisted the nozzle switch of the lamp to allow paraffin up the heating tube into the silk filament. Smoke belched out of the top of the light and drifted away on the light breeze. I flicked the switch on and off. If the mantle didn't light, I would have to start the whole process again, but luckily the old lamp started to glow. The silk mantle turned from yellow to white as the hot hissing paraffin hit. My face glowed in the reflected heat. I adjusted the lamp until it was just right and pumped up the pressure to keep it going. It hissed back at me with satisfaction. I washed my hands in the water and looked for the next hole to dig. There was no sign of Dad; I knew he wouldn't be coming as it was nearly dark. He would have my supper waiting for me when I got home. I stared into my bucket of lugworm with satisfaction, knowing that it would mean plenty of fish for us.

I emptied my bucket and I counted 106 lugworms. I had done really well as I was never a great digger. I knew another half an hour or so and I would have all of the lug I would need for days. My hands were yellow from the iodine in the lugworm and my veins were bulging along my arms from the exercise. I tilted the lamp in the sand so that it shone into the holes as I dug. I moved slowly back from the incoming tide line where the digging was easiest, and moved from one sandbank to another searching out the last of the dry sand as the water came in. The tide moves fast when it is a long one. It moved at a walking pace towards the shore and I had to be vigilant of my position. I could see the rows of lights from the houses perched on the shoreline and above them a moonless night glowed with stars. All the noises were gone. The seagulls that had been pecking up the broken tails of my lugworms had flown, the walkers disappeared, it was just the sea and me, the only noise from my spade slicing the sand. It was shining bright and polished from the work and the lamp was hissing away. At last I was done, 120 prime black lug, all gutted and waiting to be threaded onto my sandline. I threw the last one into my bucket and stood, stretching my aching back. I was on the last patch of dry sand and the gullies were filling up all around me. Time to go.

In the dark I quickly washed all the sand from my arms and spade and picked up my lamp and bucket. I started to walk to the shore, but hang on! The shore was gone. The stars were gone, the houses – gone. I was in deep mist. The late summer haze had turned into one of those deadly sea mists that come on in autumn. In the last minutes it has spread silently around me, stealing everything I knew away. I swung around, holding my lamp high. It reflected back on me in an evil yellow light, even the sea had turned the colour of the lamplight and I could see no more than a few feet ahead in all directions.

I was standing on a sandbank like a small island and the water was coming in from all sides. There were no waves crashing down to signify which way to go. I put the lamp down and moved away from it into the sea. I turned and looked. I was no more than 20 feet away and it was starting to disappear into the black mist. Normally I can see the lights along the shoreline in both directions for miles. I quickly turned my back on the light to see if I could spot anything in the darkness. Nothing! I got back to the lamp and lifted it up as the last bit of my sandbank disappeared below my feet. Because I was on high ground, each way that I walked I went into deeper water. Because there was no wind, the tide was moving silently all around me, slowly getting deeper and deeper. Normally I would be able to hear the waves of the incoming tide and walk away from them. This was a really unusual situation, an impenetrable rolling sea mist coming in after dark on a very long tide and no wind. I was stuck a quarter of a mile out on the last sandbank that was disappearing below my feet.

Panic set in, but in that panic I had to make a decision. Which way do I walk? I could not stay there for much longer. I knew that the water would be shallow for hundreds of yards in each direction. If I walked the wrong way I would be walking into the Channel and the incoming tide. I would die as surely as walking in front of a bus, and the tide would sweep me away as soon as it reached chest height and the current got a grip. Crazy! How could it happen?

Even though I was a teenager, I was already an experienced fisherman and I knew all about the dangerous sea, more than most.

There was I, alone in the pitch black. I closed my eyes and concentrated. It had to be done; I started to walk in one direction into the sea. I knew if the water did not change depth that I was walking along a sandbank parallel to the shore rather than in or out. I walked three ways and stopped. It was now or never. I needed to pick a direction and stick to it for at least 300 yards to see if it came out on the shore or if I had walked the wrong way.

I started walking one way and slowly made my way into the deeper water. At first it was only ankle deep, then to my knees. I was breathing heavily, holding my lamp up in front of me. The water got deeper and deeper but I had to keep going. If I was right the water would suddenly get shallower as I walked up the bank towards the shore. If I was wrong… best not to think about that. Suddenly the sand below my feet went to quicksand and I sunk another foot. My waders started to fill with water and I felt the sandy grit from the sea rub between my toes. I now had my lamp as high as I could to keep it out of the water. My arms were killing me as I held them above the waves. Splashes were hitting the glass of my lamp and it was hissing and spitting back at me. If my light burst I would be in the blackness, alone. The Grim Reaper was now near and I was thinking to myself 'for God's sake, just keep going'. I wrenched my feet through the mud, step by staggering step. I prodded the ground with each foot beneath the water to make sure there was no nasty drop in front of me. Now I was at the point that I had dreaded. Soaked with water almost up to my chest. This was it! Either I had picked the right way or I was another body washed up further down the coast.

I had picked the right direction. The mud beneath my feet changed back to sand and I started to rise from the sea. A few more yards and the water was back down to my knees, then ankles. I came to the ridge of seashells that always marked the foot of the last sandbank – I had made it. As I climbed out of the dark sea I must have looked like a drowned sailor back from the dead. Anyone seeing me striding out of the blackness would be horrified and would run home to tell tales of ghosts and ghouls.

On the dry sand I fell to my knees, water rushed out of my waders. I leaned forward and my forehead pressed into the sand and I started to cry. Then I laughed out loud. My hands were shaking with adrenaline and I shouted my defiance at the black sea, clambering breathless up the shingle. I sat on the end of a wooden groyne and emptied my waders, laughing almost hysterically. I was soaked to the bone but on that warm autumn night I cycled home through the mist, euphoric. Either I was just lucky or something inside of me knew instinctively the right way to turn. I had made the right decision and survived.

As a teenager I didn't give my brush with death any more thought. Most teenagers believe that they are immortal and I had survived several attempts by the Grim Reaper already. I rolled back to 266 Coast Road and put my bike, lamp, waders and spade away. I pushed the bucket of lugworm into the gravel under the house where they would be fine for the night. I ran, in my wet clothes, up the side passage and in through the back door. Dad was sitting with a glass of red wine watching a western on TV. "How did you do?"

"Oh great, Dad" I said, pulling my socks off and wringing them out behind the back door. "I have enough bait for at least two tides of the sandline. We will be eating fresh fish for a month."

"I tried to find you," said Dad, hardly taking his eyes of the telly, "but a fog set in. You're soaked," he said, suddenly concentrating on me. "Did you fall over?"

"Yes," I lied, "I tripped coming out of a rock pool."

"Ha, well, get changed and we'll do a wash tomorrow. Supper's ready."

I quickly changed into some dry clothes and we sat together in front of the telly, eating off our laps and watching the western.

I never told Dad how close I had come to being my own fish bait!

CHAPTER 8

"WHY DON'T YOU CLEAN THEM?" Isabel screamed. In one swift movement, she reached down from her writing desk, slipped off her school shoes, and hurled them at her teacher.

As I mentioned earlier after the 'gate episode', I had taken the Normans Bay route on my way to Bexhill to call on Isabel. I was to find out that as a child she had launched her shoes at her teacher. As it turned out it was the wickedest thing that she had ever done.

She was to be my last call of the morning – she had jammed her sewing machine trying to alter a pair of curtains. Isabel had lived in the small fishing village of Normans Bay as a child and she would fill in tiny details for me and paint a picture of the thriving community that was once there.

Before World War II, Normans Bay reached its peak. Tourists came and stayed. A small church blessed the fishermen as they went off to sea. The school where Isabel went had 11 children and opened every day with the teacher, Miss Jeans, travelling in from St Leonards. She was usually the only passenger to get off, unofficially, at Normans Bay, so the train driver made a special stop just for her. Mr Dodswell, who lived in the house by the railway line, was the gatekeeper for the road crossing. He would close the crossing and after the teacher stepped down he would open them again and signal the driver to continue. Miss Jeans was a firm but fair teacher and Isabel, once, just once, got into a lot of trouble with her.

"WHY DON'T YOU CLEAN THEM?" Isabel had screamed, launching her dirty shoes at her teacher. With sharp intakes of breath and loads of bulging eyes the whole class fell into an astonished silence. Miss Jeans, the head teacher, actually the only teacher at the Normans Bay School, slowly picked up the shoes with a stare that could freeze stone.

Isabel had realised the second the shoes left her hands, that she had committed the worst crime of her young life. The room was shocked into sudden silence. The teacher's eyes became icy pinpricks, "GET DOWN HERE THIS SECOND!" she howled, in a high-pitched squeal of anger. Isabel turned the shade of one of the lobsters her dad used to boil up in his pot, as she dragged her reluctant feet along the central isle between the desks of her classmates, towards her execution. It was the longest walk of her life.

All she had been trying to do was finish her class work. Isabel was the best writer in the class and proud of her writing ability, but the constant interruptions from the teacher had made something snap. Perhaps it was the cold, dark walk to school. Not far, as she lived opposite the Coast Guard Cottages, but in the dark the dirt track was full of potholes, many filled with icy water. She had not intended to step into the muddy holes, and she had not deliberately made her shoes dirty. In fact the evening before she had spent time with her dad cleaning them by the light of the crackling fire, which he had made from driftwood washed up along the shoreline.

Miss Jeans hissed out of her tight lips "Hold your hands out." Punishment was swift and she was then sent home, banished with a letter. Isabel cried her way back and found no sympathy from her mother who was hanging out the washing. She was smacked again and sent to her room. That evening when her father returned Isabel was peering out of the window. She knew he would be furious. Her father gave her a wallop and a hug and sent her to bed. Isabel had been smacked three times for the same crime. She never did anything like that again in her life and to this day she never really understood why she had snapped in the first place. The following day she apologised to Miss Jeans and all settled down in the tiny school at Normans Bay.

Things were not calm for long as Auntie Brim told her in the village shop, "War's broken out dearie. The Germans are coming. You had better tell your dad to stock up before I run out of goods. They say there is going to be rationing."

As Isabel ran from the shop she bumped into Mr Dodswell. "Bella, you're always running around like a mad thing. Slow down, girl." "The Germs are coming," she squealed, as she shot away.

"Mum, mum, war has started. The Germs are coming over the sea, they'll be here by teatime."

"What in Heaven's name are you talking about now, girl?"

"It's true, Auntie Brim told me. She said we have to buy up all the biscuits and sweets before the rats steal them."

"Isabel, calm down, you can hardly breathe. I will pop along to the shop and see what that daft Alice Ribblesford is gossiping about. Now you get in and do your homework."

CHAPTER 9

Auntie Brim was right, World War Two had started and their lives would never be the same. At first there was little evidence of any war. A few extra children turned up in Eastbourne late in 1939 but were soon ferried away to rural villages. The 'Phoney War' rumbled on with a few false air raid sirens scaring everyone. The houses were blacked out, street lighting turned off and 1940 rolled in as quiet as a church mouse.

All carried on as usual, the fishermen, the milkman and the latest gossip, all in equal amounts. Rationing started to bite because far away in the cold North Atlantic the Germans were destroying millions of tonnes of shipping. To keep up morale, news was censored and the papers told little of the terror and carnage that was happening far away. Locals turned over their flowerbeds and planted vegetables, and everyone scoured the papers for the latest news. Normans Bay was a highly likely spot for the German invasion and Customs and Excise men were re-trained as lookouts.

In the March of 1940, fisherman came back excited with the news that a merchant ship had been bombed off Beachy Head. The SS

Barnhill had been seriously damaged and the Eastbourne Lifeboat had rushed to its aid. Several of the crew were already dead but 29 more were saved as the ship turned into a fireball.

After the ship was cleared, a bell was heard ringing back on board, so local fishermen Alec Huggett and Tom Allchorn jumped back on the stricken vessel. They found the ship's Captain, Michael O'Neill, badly wounded. He had refused to leave until all of his men had been accounted for. While searching the ship he was badly injured by an explosion. He managed to crawl along the burning deck to pull a bell rope, ringing it with his teeth. They lifted him to safety aboard the Jane Holland Lifeboat. The SS Barnhill continued to blaze and drift for days before running aground just east of Langney Point. It slowly died in front of crowds that had gathered to watch the spectacle. In a final twist, its main funnel dropped into the sea and the ship broke its back. Black smoke billowed out of it as the saltwater rushed in. However the ship had one last surprise.

On the high tides, out of the broken hull, floated all sorts of goodies. Some were edible some were not. Typewriters rusted and the boxes of spare ribbons were ruined, the carbon paper rendered useless. There was also Dutch cheese and more. Tins of food floated and many more were helped out of the low tides by the locals. As rumours of the bounty spread people came from far and wide to grab what they could. Trails of people could be seen walking across the barren land known as The Crumbles, timing their arrival for low tide. Eventually the customs men took charge and started to threaten prosecution of anyone taking goods. They confiscated and rounded up what they could and chased away the spectators. The crowds eventually dispersed and what had become a local pastime died out as quickly as it had started.

However, on the high tides cans started to float in. Some contained meat, some peaches, some tinned tomatoes. All had lost their labels and a few were inedible, burnt black by the heat. Locals welcomed these little gifts. Cyril and Iris Cottington would sit down to supper and on the table would be a Barnhill can. Cyril would carefully open the tin that would accompany their meal. Tomatoes were nice but peaches were better. They named these suppers, 'tin surprise' and looked forward to their free bounty.

"Open the door Bella, quick girl." Isabel quickly opened the back door as her father struggled in with a large wooden box. He stared back out through the door to see if anyone was watching and quickly locked it. "Look what has just drifted in on the tide," he said breathlessly. "It's a truckle of cheese from the Barnhill. It's come in on the night tide like a gift from the gods. There's a load of them so we shared them out. Be quick now, fetch your mother then bring the spade." Isabel ran off with excitement while her father went about opening his prize. By the time Isabel returned with mum a huge cheese was sitting on the dining room table. "What about that beauty, eh! More cheese than we have ever seen, and not a penny paid." The family danced around the table with joy. Percy then lit the fire and started to burn the packaging as quickly as he could. It was still wet so he was using his old, dry wood and dropping bits of the wet stuff on top. "Bella, you keep topping up the fire. I'm taking the spade to the garden to bury the cheese. The customs men will be here as soon as they hear about the wash-up and they ain't getting this beauty without a struggle."

Later on Isabel went outside to find her father. He had dug a hole going under their little pond. As soon as it was large enough he wrapped the cheese in his fishing oilskin and buried it. He then hid all traces of his digging. Isabel held up the old lamp, her eyes glowing large with excitement. "I can't see a thing Dad, I am sure that no one will find it. Mum's burnt most of the cheese wood."

"Good. Time for supper, I'm so hungry I could eat you all up." Isabel squealed and ran for the door as her father chased up the path. The next morning, sure enough, along came the customs men. One by one they pulled out the cheeses from the other fishermen's houses. Some were hidden in lofts, some under the upturned boats and outhouses. One was in the loo down the bottom of one of the gardens. No one spoke to the customs men. The fisherman and customs men were historic enemies. Eventually they left and the families were allowed back into their homes to clear up. "Can we go and see if they have found our cheese, Dad?"

"Not yet, girl" replied her father, peering out of the window. "We'll leave it for today just in case they are watching. I'll pop out after dark and see." It took an age for the sun to drop and eventually Isabel's father slipped quietly out of the back door. She sat impatiently waiting, drumming her fingers on the dining room table. She was dying to know if the cheese had been taken. Eventually Percy came back in with a sad expression on his face. His hands were empty and dirty. He walked over to the table rubbing his head. He looked at Isabel with his forlorn expression and then reached inside his shirt. He pulled out a huge slab of fresh cheese and put it on the table. Everyone cheered and Isabel clapped and clapped till her fingers hurt. That night Isabel toasted bread in front of the fire and ate cheese on toast until her belly hurt. The next day the fishermen shared out all the cheeses that had not been found, and over the next few weeks Isabel had more cheese on more food than she had ever eaten in her life.

Life in Normans Bay took a turn for the worse when notification came that every house had to be evacuated. Families had to make their own arrangements to stay with relatives or friends but anything left by the end of the week would become Government property. It was the final death knell for the old village. Before long the beaches were wired up and mined. Anti-aircraft guns lined the coast and the heavy German bombardment of the South Coast towns started. The Phoney War had ended and peace was shattered.

CHAPTER 10

At the end of the war as the final all-clear sounded to signal the end of hostilities, women dragged down their blackout curtains and threw open their windows to welcome a brave new world. Unfortunately, like many small towns and villages the war had changed everything at Normans Bay. Families that had left never returned, buildings were demolished, some left empty and the small fishing hamlet became a ghost of itself, left to the elements.

As the decades rolled by holidaymakers still came down and camped at the local sites. They played on the open beaches, cleared of mines and wire, and made sand castles. But there were few of the local families left that could tell the tales of old. Isabel would no longer get a penny for standing with a holiday maker for a photo and then run off to buy an ice cream at Auntie Brim's. The café opened for a time but closed again. The church, the school and the community all disappeared.

Occasionally I stand with my back to the sea and look at the small hamlet. As the copper sun sets in the west, as its last rays repaint the peeling houses in a new coat of burning orange I see how it once was. I hear the noises from the busy little school as Miss Jeans rings the bell for the end of the day's lessons. I see the children running home as their fathers are pulling their boats up the shingle bank. I see the women gossiping in their gardens, the café full of holidaymakers. Auntie Brim has a blackboard sign out with her special offers written in chalk.

Now they are all ghosts. A small, proud and independent community still survives there, amongst the empty holiday homes. However, a long time ago, the tiny hamlet of Normans Bay had thrived and a little girl lived on cheese hidden from the excise men.

SPRINGTIME BLUES

Blossom bursting forth this Spring.
Sweetness from the birds on wing.
Sorrow in my broken heart.
Woeful pain and eyes that smart.

Lambing ewes in folds of green.
Primrose bloom, the ideal scene.
Nothing helps to ease my woe.
Moody days where cobwebs grow.

Skylarks sing and the doves pair.
Locals busy with the Fayre.
Kicking dirt I shuffle by.
Misty thoughts that make me cry.

Bluebells coat the forest floor.
Farmers joke as years before.
My love has gone how I pine.
Lonely days are my springtime.

LEFT: This is Mumsie at the Ripe Fayre, I think in 1964, I was very young, hiding in the tent as she read palms. Her soft Austrian accent and her glamorous looks made her the perfect person.

BELOW: Herstmonceux castle is hidden away close to Eastbourne and many don't bother to visit it, but the castle is one of the hidden gems of Sussex.

Here is Yana in my sewing machine museum. She is holding my favourite sewing machine called a Moldacot. This particular model was sold in Whitechapel, London, around 1888, exactly the same time and place as Jack the Ripper.

I spend countless hours in my four workshops, building and repairing sewing machines. It is lucky that my hobby and my job are one and the same. If no one knows where I am – they have to search them all.

LEFT: This is one of the only pictures to survive of Nelson Reed. He went off to war, aged 19, and died in Flanders. From the story, 'The day That Sussex Died'.

BELOW: This is the Little Common to Normans Bay sea-road. I am driving along what is left of the road during a violent storm which has disappeared under shingle.

ABOVE: My old Daimler, which I have had for over 35 years is still going strong. I often use it for friends' weddings, and here she is outside Herstmonceux Castle with the bride and groom, Tracey and Glyn Ripley.

LEFT: This is my daughter, Sarah, on the downs in autumn of 2012. The colour carpeting the hills was breath-taking, and although she hates her photo being taken (typical photographer), the scene was just perfect.

Still our first line of defence, the ancient sport of archery is as popular today as it was centuries ago, and clubs meet most weekends all over the country. Here we are at Scotney Castle celebrating Armistice Day 2012.

Bodiam Castle is a photographer's dream — and they have great tea-rooms. When I am out on calls, I often pop in for a cup of coffee, the place is magical.

Here I am with a small part of my antique sewing machine collection which has grown for decades. I now have the collection that I have always dreamed of. The youngest machine dates from 1880 and the oldest 1859.

This is a fascinating picture. People assume that most sewing machines are similar. Because of patent protection, nothing could be further from the truth. Here I am standing next to an Italian Dueffe SM300 Quilter, the size of a large room, but I am also holding a sewing machine. If I can't fix the Dueffe they have to fly an engineer over from the Italian factory, and it costs them a fortune!

Director at Advanta, Mike Cottingham, presents a certificate to Yana Askaroff

In between working, helping me and normal family life, Yana works tirelessly raising money for charity and every now and again she gets acknowledged. Here she is in the local paper receiving an award.

St John's Church in Piddinghoe. The church is an unusual shape perched on the edge of the River Ouse. I was told that East Sussex has more flint churches than anywhere else in the world.

Pevensey Castle is just an ancient ruin now, home for pigeons and field mice but a wonderfully atmospheric place to walk around on a wild day, or at night if you are brave enough.

This is Eastbourne beach by the pier, being topped up. The beach is artificial and millions of tonnes of pebbles are sprayed onto the beach to keep it high. When I was a kid, the beach was much lower and had loads of sand.

Two iconic images, my old 1966 Daimler and the Long Man of Wilmington in the background. I always think that my Daimler is so reliable because it was built the year England won the World Cup. I often imagine that the workforce were happy when they built my old beauty and that is why, in over 30 years, she has never failed to get me home.

Here I am at J E Smith Upholsterers. I used to repair the machines for old man Smith, but now his sons Jim and Gary, run the business in Eastbourne. The machine I am fixing is a Solent Long Arm, and it is so large you could slip through the gap between the table and arm. It is used for large upholstery and sail cloth. It has hydraulics and pneumatics controlling the operations and is a tricky little blighter to fix.

Tina's Pizza

THE RARY

At my boarding school we spent our evenings sitting on our beds, some of us doing homework, others projects or chatting. The teachers took it in turns to stay late for lights-out at nine. Of all the teachers, Mr Cousins stood out. He was our science teacher. He was always supportive, encouraging and interesting, the perfect teacher. Often before lights-out, if we pleaded and sounded desperate enough he would occasionally weaken and tell us a story. We would all sit patiently on our beds as he would tell us one of his amazing tales, and at the end, he would say goodnight and turn the lights out.

Of all the stories he told us, one stands out. We laughed so hard that he had to come back upstairs twice, to tell us to be quiet before we woke up the other dorms. I am pretty sure he made it up. Here it is:

One day he was walking along the cliffs at night with his dog when he heard something. He looked round, and coming over the edge of the cliff was this small creature. He was dumbstruck and watched in amazement as the little fat creature stood up and dusted himself off. He looked at Mr Cousins and smiled with a charming smile. "Hello," said Mr Cousins, as his dog sniffs the little fellow. "Who are you?"

"I am a Rary." Said the little fellow. "Have you any food?"

"Yes, I have some back at the school. Come along and I'll see if cook can make you some supper."

Mr Cousins, his dog and the Rary walked back along the cliff path to St Bede's where he got cook to make up some grub for the Rary. He tucked the little monster up in one of the spare school beds and went home. When he got back to school the next morning there was pandemonium. The school larder had been broken into and lots of the food had gone. The children had to have bread and water for breakfast. Mr Cousins had the sneaky suspicion that it had something to do with his new friend so he went to find the Rary. He found the monster in his bed, as fat as a barrel with crumbs all over him. He was snoring like a pig. Mr Cousins woke him up and demanded to know what he had done but the monster denied everything and charmed his way back into Mr Cousins' good books by cleaning his spare school shoes.

As the days passed food kept going missing, even the tuck boxes were broken into. Each time Mr Cousins suspected it was the Rary but he had no proof, and each time he confronted the Rary, he was so charming and smiled so sweetly that Mr Cousins forgave him – that was until the headmaster's wife's afternoon tea party. Mrs Pyemont was having an afternoon tea party for some of her friends and she had bought lots of lovely sandwiches and cream cakes. When her friends arrived they went into the room to find nothing but crumbs. You could hear her screams all around the school as she furiously searched for the culprits.

Mr Cousins knew who had eaten them, and rushed to the Rary's room to find him stuffing the last cream cake down his gob. As much as the Rary pleaded and squirmed, Mr Cousins told him he had to go. However, the Rary had a good life at the school and was not budging, so Mr Cousins had to come up with a plan. He decided to wait until nightfall and take the Rary back up to the cliffs to where he had found him and then let him go. Mr Cousins chased the fat little monster around the room until he leapt on him and pinned him to the floor. As the Rary wiggled he tied him up with school ties and then he gagged him, as the monster was very persuasive and he did not want to change his mind. After dark he put the Rary in a wheelbarrow and pushed him back up the long cliff path that ran up by the side of the school.

It was a dark and stormy night when Mr Cousins arrived at the cliff edge, at exactly the same spot where he had found the little monster two weeks before. He felt a little sorry for the Rary but he had to go. He knew he would never see the Rary again, but now he always smiles when he thinks of his last words.

He untied the Rary and took his gag off. "What are you going to do with me?" Asked the Rary.

"I am going to chuck you back over the cliff, back to where you came from."

"Well I say," said the Rary laughing, "That's very rude of you. Would you mind if I just peeked over the cliff to see how high it is?" As Mr Cousins pushed the wheelbarrow closer to the edge, the Rary peeked over the edge, then, as quick as a flash, the fat little monster leapt out of the wheelbarrow and disappeared into the bushes, laughing as he went.

All Mr Cousins could hear in the dark was the little monster singing as he disappeared forever into the night,

"THAT'S A LONG WAY TO TIPPARARY,
THAT'S A LONG WAY DOWN THERE."

THE CHRISTMAS GOOSE

Pans are all a-sizzle pots are getting warm,
Mother's in the kitchen cooking up a storm.
Father's in the yard the chopper's in his hand,
The goose is all a-flap for a goose dinner's planned.

Blubber boy's on the run – no longer getting fat,
Chopping blocks ready his heads a-going splat.
The dog is all-a-fluster barking at the sky,
Roast on Sunday makes next week's pie.

Mother's in the kitchen cooking up a storm,
Pans are all a-sizzle Pots are getting warm.
Father's in the yard his hand is on his head,
Blubber's flown the coupe but he should be dead!

He'll have to have to tell the wife, the goose hasn't died,
He had better hide that chopper before he goes inside!

TINA'S OLD NAPLES' PIZZA

When I was a kid we never went to restaurants. I am not alone in this, it seems to be my generation. Our parents went out, we stayed at home. Besides fish'n'chips bought with money earned, my first memories of restaurants started in my late teenage years. The kids today have no idea how lucky they are with all the eating places available to them on a regular basis.

The reason I am telling you this is because while growing up our daily diet was pretty boring; we grabbed what we could most days and on Sundays, a glorious meal would be presented before us. This routine was broken by occasional delights, and Tina's pizza was one of them. Now Tina's pizza was not just good – it was amazing. Grown men would happily go to their graves for a single bite. Lemmings would leap over Beachy Head following its scent. Their little bodies, prostrate on the salty rocks, were miraculously raised from the dead by wafting a slice of it under their noses. Well, believe it or not, we don't have any lemmings in Eastbourne, but you get the idea. Her pizza was just amazing, and the sight of it coming through the door in the huge tray that Tina used, would make Scrooge smile. As I am typing I can feel myself drooling. Her pizza was famous in our little corner of the world, and she managed to get just about anything she wanted from us by using it as bait.

I am going to give you the recipe for her traditional family pizza from Naples; thick, filling and absolutely gorgeous. It is more like a focaccia bread than the typical pizza that you see today, but this really is an old family pizza combining the best of both. The well-guarded recipe took me years to gain but it was well worth it. First, let me explain a little of how I eventually managed to procure it from my old Italian friend.

Tina Caccavrle was born in Cicciano, outside Naples just before World War II. She had four sisters and seven brothers, 12 in all. At dinner the family of 14 sat down to eat along the rough wood table in their large, dilapidated Naples home. Getting everyone fed at meal times, and lunches packed for the following day, especially during WW2, was always a top priority. They needed cheap, delicious food.

Tina often helped her mother prepare their meals and this family recipe has been passed down for generations. The family tradition goes that one of her distant relatives made too much dough mixture for her small amount of mixed herb, ham and vegetable topping, so she made the base much thicker than normal and the family loved it. Tina learnt this unique recipe and boasted that the Roman army marched on it.

There is also a legend in Naples that the local prostitutes made several quick and tasty dishes for visiting sailors. These tasty treats kept the men satisfied in more ways than one! The girls could quickly prepare the food in their bedsits and the smell was so enticing that the sailors would wait for it. Apparently happy energetic sailors were good for business!

Tina came to England for a short break and ended up staying forever. As a young girl she found love, but I will explain that in a moment. Tina and her husband Ronnie Englert were both working at our family firm, Simplantex, where I grew up. She used to make my parents, and later me, these fabulous pizzas every time she needed anything. I don't think I have ever received payment in money from Tina for fixing her sewing machines! Her pizza was so good that it was an easy bribe.

Ronnie was a red-haired monster who ran the cutting room at the family business. He had a deformed hand, or so I thought. As I grew and started work in the cutting room, I saw something amazing. As he slipped his huge tailor's shears onto his cutting hand they fitted his weird shape. His hand and muscles had taken the shape of the scissors that he had used for decades. As I got older I assumed that because of the way he acted, although he was married with two boys, he was probably gay or as Roy, our sensitive van driver once said, "As bent as a three bob note." We always got on like a house on fire.

Ronnie was small and muscular, with a shock of red hair and a rash of red freckles. He had a wicked temper and a short fuse, to match his stature. For no apparent reason he would blow his top, have a five minute tantrum, and then be all smiles. The girls in the sewing and packing rooms would steer well clear of him when the full moon came around each month.

I almost grew up in the cutting room, where all the fabrics were laid out each day for the various items on our price list. I worked with Ronnie for years and used to watch him prance home each night with his upright wiggle. Even his bicycle, which he occasionally used, was a ladies' green Raleigh with no crossbar. He only lived two roads down from our factory in Willowfield Road, and would sometimes creep through the back window of the factory and down an alley to his house. I got on so well with Ronnie that, when he was in the mood, he would talk like no man could talk; we would get lost in time as he chatted away. Quite often we would be the last to leave the factory and I would walk to the gates with him to lock up the yard. One day he gave me his old tailors shears, (I still treasure them and use them most days). "They are too heavy for me now, you might as well have them. You've learnt enough to handle them. The blades are sharp and the hinges oiled. I have just had them serviced and set by the ironmongers, Pittman's, down Cavendish Place. They were made by Jacob Wiss in America, cold-forged, hollow ground, tempered high-carbon steel. They are a work of art in engineering and will slice through a single thread of a cobweb, a sheet of finest silk or the hardest denim. I bought them with my first month's wages as a cutter and used them everyday at work. One day you can pass them on to your kids and tell them where they came from."

It was during one of these chats that he told me how he first met Tina. Years later, Tina told me the other side of the story, of how she first met Ronnie. It was wonderful to hear both sides of the same story. I always had the sneaking suspicion that, like the rest of us, she had trapped him with her pizza!

Tina used to sew up some of the products that we produced at the business, working at home as an outworker while she brought up their kids. They both worked at the business and as the years went by, we were bribed with pizzas on a regular basis. I can't tell you how they made my day. I would so look forward to my pizza and, although I would stuff my face, I would feel great for days.

After I left the family firm I still looked after Tina's sewing machines and loved the call from Tina when she needed something, as I knew what my reward would be.

Years passed by, and one day Tina and I were chatting and I asked about Ronnie who had died several years before – it was then that I told Tina of Ronnie's version of their meeting. At 73, she learnt something from years before that brought a tear to her eye. And I learnt why Ronnie often behaved the way he did.

Before Ronnie worked at Simplantex he was working at one of the big seafront hotels in the 1960's. On regular nights, like many prominent buildings in town, dances were held. 'Real' dancing was unbelievably popular, from the 1930's right through to the 1970's, when it fizzled away, only to come back a little after the Millennium. One particular dance-night was Ronnie's night off, but he had nothing to do and decided to get a drink at the bar of the hotel. Fate was about to change his life.

As he walked past the dance floor on his way to the bar, he casually glanced across the room. In an instant he was frozen to the spot. In the middle of the floor, a black-eyed beauty was dancing. It was Tina, fresh from Italy wearing a new skin-tight dress that she had just made. Ronnie told me that it was love at first sight, he just could not keep his eyes off her. He felt his heart beating so hard he thought it was going to burst out of his chest. He instantly knew she was the one; his girl, his wife, the mother of his children. The woman he would cherish until his dying day!

He rushed back to his room and changed into his best dance gear. He then rushed back, and pushed his way through the dance floor to where the tanned girl with the long black hair was dancing with her friend.

Tina then told me something that made sense of the way Ronnie was. He was a semi-professional ballroom dancer. I never knew. Of course he was – the way he held himself, his stature, muscular physic, prancing walk, perfect little ways. He wasn't gay, he was a dancer! Doh. Tina telling me this was a revelation and a lifetime of misconceptions evaporated in an instant.

Tina never knew that Ronnie had rushed back to his room to change on that night. It was the first time that she had heard Ronnie's version of their meeting. He had played it cool, dancing away the evening on the dance floor with his new Italian lover. What he lost in height, he made up for in pure style. As well as the classic dances, Ronnie showed her The Bristol Stomp, The Madison, The Continental, and more. Tina was blown away by his dancing ability and even more by his amazing red hair. She had never seen anyone like him; a little ball of thunder.

At the end of the evening Tina was besotted, and when Ronnie hesitated she grabbed him and showed him how Italian girls kiss. Tina never went home to Italy – from that night they stayed together, through thick and thin, for life.

I remember Tina so clearly after our long conversation on that lazy afternoon. We both sat in silence for a while, sipping tea and savouring the tête-à-tête that we just had, almost living it anew. Oh, how I would have loved to have seen that beautiful young Italian on the dance floor that night, when Ronnie introduced himself and showed her how he could shake a leg. What I would give to witness real love at first sight.

Anyway, that was long ago and, as usual, I have digressed away from my original purpose of telling you how to make Tina's amazing Naples' Pizza.

After at least 35 years of pestering and begging Tina, at 73, and in a moment of weakness, she finally gave in and gave me her family recipe. It is a real taste of heaven.

You don't have to be exact with everything, just imagine Tina mixing all her ingredients on a big old table in Naples. Oh, after Tina gave me this recipe I made it almost every week. I put on half a stone and ran around like a kid with endless energy. I had to limit myself to making it once a month. Don't blame me if you get fat!

This recipe will feed about six (or three of me). It is amazingly quick and easy to make. The smell is wonderful and I can just imagine the Naples' working girls preparing this fast food.

The flavours that topped the pizza varied depended on what leftover's the family had, or what vegetables were preserved, tinned, bottled or in season. The base flavour of the pizza is always the same, olives, garlic and herbs which were available all year round.

Cheese was too expensive and hardly ever added but feel free to pop some on.

The main body of the pizza

18oz Plain or bread flour.
7g Dry yeast.
10 Fluid oz water.
10 Garlic cloves, (five for the base).
2 Chillies, one for the base.
10 Olives, any type, five for the base, more if you love them.
4 Sun dried tomatoes (two for the base), more if you love them.
1 pinch of Mediterranean herbs.
Olive oil.
1 Pinch of salt and pepper.
1 Teaspoon of sugar.
4 Tinned tomatoes, (they will be mashed up).
8 Anchovies.

For the toppings you can mix and match your favourite flavours. Over the years Tina added, mushrooms, ham, roasted peppers, bottled/tinned mushrooms, Parma Ham, fried bacon, salami and even capers. There were never two pizzas toppings the same, but all the different toppings were always on her magical base. Leave out anything you don't fancy as they are all optional, so if you are a vegetarian or allergic to any of the ingredients it's easy.

This is the best bit. Take your watch off and roll up your sleeves. Get a large mixing bowl. You can use a mixer if you prefer. Mix the flour with the yeast and one cup, (about 1/2 pint), of lukewarm (not hot) water. Add one big pinch of mixed Mediterranean herbs (oregano, marjoram, rosemary, parsley, etc) and a pinch of salt, teaspoon of sugar (it feeds the yeast) and a good wallop of cracked black pepper. Stir with the spoon until you can't move it and then get your fingers in there. Knead or mix into a silky-smooth stretchy dough ball (it's a real workout). Add a touch more water or flour if needed to get the perfect stretchy consistency but don't make it too wet. It is hard and enjoyable work and you can see why those Mediterranean women can crack walnuts in their hands! You will be able to feel the springy dough when it is ready. This usually takes about 10 minutes.

Chop up and mix in the five garlic cloves, one or more chilli's, two or more sun dried tomatoes and five or more olives of any colour. By now the kitchen will be smelling like an Italian restaurant. I have even been known to put Dean Martin on in the background singing *Volare*!

Oil the base of a shallow tray of about 12 inches with a healthy dollop of olive oil which soaks into the pizza and crisps up the base. It is important that you use enough or it may stick. Tina always used a large roasting tin which works well, but make sure you use plenty of oil. Olive oil is a miracle food anyway.

Roll, press and knead the dough into the tray to roughly ½" thick. This is not as easy as you may think for the dough is like elastic and doesn't want to reach the corners but keep at it and remember, this is a fabulously thick, yummy pizza. You can use a larger tray to make a thinner pizza. Cover and let it prove somewhere warm until it rises and doubles in size, which is usually about an hour or so. Do not put it in the airing cupboard like I did unless you take all the other stuff out first. I didn't, and my fresh sheets stank of garlic!

While the base is proving you can chop up the rest of your ingredients for the topping and clean up. I love the cleaning up of the garlic skins and jars of goodies. If you don't like the smell of garlic on your hands, wear surgical gloves. Did you know that if you rub garlic into you hands, your breath smells of it the next day?

Once the base has risen, prick the pizza all over the top with a fork and spread more olive oil on. Mash up the anchovy fillets into a paste and spread evenly over the pizza, it adds a real bite. You cannot tell that they are anchovies, they are really more of a seasoning like soy or fish sauce and they add a real Mediterranean zest. Yana insists that she does not like anchovies so I spread them on my half.

I cannot stress enough to make the topping tasty as it has to compliment the thick base. Mash up the tinned tomatoes until they are almost like a lumpy puree and spread evenly the top. Add the rest of the olives, garlic and bits to the top. This is where to add all your favourite extra toppings. You can also add a little chilli oil or chilli flakes if you like heat, and sprinkle generously with olive oil, especially the corners. Remember the larger the tray the thinner the end result, so spread the toppings out accordingly. It needs to be a thick and tasty topping.

The pizza will keep for hours, so about 30 minutes before you want to eat, place the pizza just above the centre of the oven for 20-30 mins on high, gas mark 8, 230c, 450f. Keep an eye on the pizza as it cooks really quickly that high in the oven. Remove when the edges brown and crisp. The pizza should rise to at least one inch thick, hopefully more. Carefully remove from the oven and loosen the pizza with a plastic scraper. Tap the base, if it is not crispy make sure there is oil underneath and put over a gentle heat on the hob. It quickly crisps up. Then let the pizza cool, just a little. A crispy base is essential.

Place on a cutting board and slice. Serve warm in chunky slabs with a fresh salad and a glass of your favourite tipple, or keep it as a cold hearty snack which will lasts for days. There you have it. One of the tastiest foods ever to grace this earth. Thank you Tina.

Sit back and enjoy a real taste of old Napoli and imagine two young lovers in 1960's Eastbourne, coming together for the first time across the dance floor.

HALF WAY TO HEAVEN

Isabel was fretting in the kitchen, "They're going to kill them all! I know they are."

"Kill who?" I asked with a worried expression.

"The cockerels...all 17 of them," she said, pacing up and down and constantly staring out the back door, to where a car had brought the executioner. "I just hope they don't do Darcey. He's my favourite, and Chunk."

"Chunk?"

"Yes Chunk. His mother pecked his neck when he was a few days old and took a chunk out of it, now he flops around with a wobbly head. It is really sweet you know."

"I'll believe you. So, tell me why are they all going to the block today."

"Oh, it's Mum, she can't stand the racket. It is pretty bad really, it starts around 3·30 each morning. The noise is amazing, 17 cockerels all 'cockrelling' or whatever you call it. It is lucky we live on a farm or our neighbours would have shot the lot of us by now. I think I am going to have to see what is going on, I can't stand it any longer."

With that, Isabel shot out the door and ran up the path towards the back of a large farm building where she told me all the 'action' was taking place. I settled down to repairing her Singer which had taken off on its own, and was trying to sew by itself every time it was plugged in. I kept looking out the back door wanting to know what was going on. I wanted to be up there to see what was happening. In my mind's eye I could see Isabel arguing with her dad.

I stopped for a break after repairing the Singer and before I started on a Toyota. I picked up my coffee from the old farmhouse table and leaned against the door frame of the rickety old building that was straight out of some Dick King-Smith novel.

Mum, Alison, had made me a perfect cup of coffee, with two sugars now that I had given up my diet. When she had asked me how I liked my coffee, I told her with bacon and eggs! It was worth a try. I had realised in my long travels that there were many things in this world that were trying to kill me, but an extra spoonful of sugar in my coffee was not the one most likely to succeed.

I was alone in the house. While I leaned against the door frame, I crossed one leg over the other, took a sip of coffee and soaked up the day. House martins darted around gossiping in short 'chirps' as they caught insects and flew to the barn, where I guessed their young were waiting with huge orange open gobs in their cosy mud, twig and feather nests. A couple of grubby farm dogs, possibly brown and white spaniels (it was difficult to tell under the mud), lay exhausted in the dusty track which bumped and twisted for a quarter of a mile up to the main road. The air was warm and sweet with the smell of fresh cut grass. I could hear a tractor working a field somewhere out of sight.

In the yard, a string of fresh washing swung in the breeze, as the trees in the woods behind moved gently in the summer air. Old bits of disused farm equipment lay around and a dented milk churn leaned at a precarious angle in the flowery meadow-grass. High above, a jet-liner cut a line across the heavens and way above that, the clear blue sky sparkled all the way to infinity. Moments captured like this are priceless and I suck up each and every one.

Suddenly, I saw Isabel running at a pace towards the outbuildings near the woods. Jammed underneath each arm was a cockerel. She hot-footed it out of sight as I laughed. 'At least she saved two of them' I thought.

I spotted Alison, her Mum, striding down the path in a very business-like manner. She was wearing blue jeans and wellies, her loose brown gingham shirt had the sleeves rolled up and her hair was blowing in the breeze. She was as pretty as a picture. "Well, we have done it," she sighed. "We have dispatched nearly all the cockerels. Isabel intervened and saved her two favourites. Darcey will live for another day. Although I imagine now that the numbers are down to just two, the local fox family will have a beady eye out for them. Oh isn't it a glorious day and so lovely to get the washing out?"

"Yes," I said. "There is something quite special about looking out of the window and seeing that the sky is clear so the washing can get an airing."

"Alex, anyone would think you were a woman!"

"Right," I laughed. "The only way I would get away with being a woman is if I joined the 1970's Russian shot-put squad. I am not sure if I would need to shave though! Actually it is based on experience. I have about 30,000 women customers so I have learnt a few things over the last few decades and hanging fresh washing out is just one of them. In fact I often wonder in my unique job, growing up in a female dominated trade, if I know more about women than any man alive. But you just have to tell me, why on earth have you so many cockerels? Most farms have just one."

"Oh, it's the ballet school in Lewes where my daughters go."

"Yes of course," I laughed. "Why didn't I realise that the ballet school in Lewes handed out cockerels to its pupils. Although it does explain Darcey, named after the ballet dancer Darcey Bussle, even though it's a boy."

"I know you think I am mad, but at the end of term the ballet school give each pupil an egg to hatch. It is a sort of discipline. I think that the idea is to set them a task which they have to follow rigorously or the chicken dies. Anyway at the end there are always loads of cockerels that no one wants and, as we have a farm, we end up with the extra cockerels each year. At first it was not too bad but as time went on, more and more of the little devils arrived, running amuck on our farm. We had to make a decision. So we called in the executioner, and today was the last day for most of them."

I looked on in the simple realisation, that farm life, is life and death, on a daily basis. While her daughters would take it to heart, both parents had been amazingly patient to allow the numbers to get so high in the first place. Cockerels out-competing each other at three-in-the-morning, crowing their little hearts to out announce the new day would drive most people crazy. I know I would have been up the track with an axe long before now!

I was back at work busily attacking Alison's Toyota sewing machine when a forlorn looking Isabel arrived. Her eyes were deep and dark as if she had been crying and her short shiny back hair had a piece of straw in it. I just smiled quietly knowing the enormity of the situation for her. "They didn't get Darcey," she said with a pout of defiance. "it is like something from a horror story up there, I think I'm going veggie from now on!" With that, she stamped upstairs.

Her Mum whispered, "Won't last a week, she loves her bacon too much."

As I wound up the path, I took one look back at the farm and wondered if next term they would still take the unwanted cockerels!

The day had been one of those wonderful days where I would have worked for nothing. First call was at Worlds End, a village on the Sussex border now almost swallowed up by Burgess Hill, next onto an old convent still run by a few nuns called Holy Cross Priory, near Heathfield.

The Priory had been one of the many that had been modernised by a firm from the West Country. They specialise in 'doing up' these old mansions in the middle of nowhere. The priory itself was carefully brought back to its old impressive retro-gothic grandeur, straight out of a Dracula movie. All around were new flats for people who want a little patch of peace away from the hustle and bustle of town life.

Janet had been at the old convent flats for years before the renovation and had now moved into new apartments overlooking the bowling green. "I am not sure if I prefer the old flat or the new?" Said Janet offering me a biscuit. I knew I would be stopping at the Albion Bakery near Cross-in-Hand for one of their amazing apple turnovers so I politely declined. "The old flats were larger but my new one has a better view. I think once they start playing bowls I will be glued to the window. I have already had my electric organ moved there so that I can look out while I am playing."

Janet was 86 and still going strong. "Why do you think I moved here Alex?"

"Well I guess you believe in God and moved to the convent to be closer to him. Oh, and I know you used to love that huge crimson rhododendron that flowered outside your old flat"

"Yes and no. I moved here because this part of the world is so beautiful, I always think of it as half way to heaven and I decided that I would fight my last fight here."

"Expecting to be attacked are you?" I laughed.

"Of course Alex, after a certain age we get attacked on a daily basis. You have to look at the bigger picture, life on a cosmic scale. We are connected in this universe. Our bodies are a cake mix of everything around us. All that you can see we are part of. It is our immortality.

Some people call this extraordinary mix Mother Nature. Mother Nature needs us as much as we need her. Our world is supplied with everything we need but there is a price to pay and we pay it with our bodies. You see Mother Nature uses us to fertilise her soil. Think about it for a second, billions of creatures have been living and dying on this planet for thousands of millions of years. The planet is constantly regenerating. You don't have to believe in God to see it, it's as plain as day. I am little use anymore and so my body will soon become one with everything around. Part of me will help generate the next living creatures on Earth.

Mother Nature is beautiful but she has a brutal beauty and by her very nature has to be practical. My cycle is nearly over and I fight each day to stay a little longer. When I give up she will take me and use me, just like a farmer takes his fertiliser to help the following season's food. That is the way it is, and the way it always has been and there is nothing that anyone can do. Not since the birth of the human race has one mortal being managed to wriggle out of their destiny. Now our souls on the other hand, well that is another matter and can be rather more heated, so I won't try and convert you today. Not until you have finished my Janome sewing machine," she said smiling with a wicked grin.

It had been like that all morning, one great call after another, I thought as I drove the country lanes of Sussex that were all coated in the beautiful summer's day. I was on my way from the cockerel incident at Gun Hill down to Apple Tree Cottage in Hailsham for a gear replacement on a Singer 507. Then off to Lover's Farmhouse in Chalvington to resurrect a seized machine. My belly and heart were full and no therapists couch on earth could have done more to resurrect the spirit than this simple morning had done. Mind you I could have stuffed another apple turnover down! Such will power...

THE SUPPRESSOR MAN COMETH

Geese were nosily forming a dart in the sky over the City of Brighton, and great fat clouds with dark-heavy bellies, wobbled their way south towards the sea, which was as grey as a Welsh slate mine. They cast large shadows over the city below as they broke the clear blue September sky. The cool air announced that autumn would soon be knocking on summer's door and the season of short dark days and long cold nights would follow. Another year was flying away.

Brighton is one of my favourite places in the world. Twenty four hours a day it is always exciting and full of interest and colour. You could write a thousand books on the city and still never cover it all, from kings and mistresses, to gamblers and fishermen. All human life is in Brighton. Of all the stories there is one that is old, almost forgotten and amazing for the history of our country turned on it.

In the graveyard of St Nicholas' Church there is a very worn and aged grave, possibly one of the oldest in the area and very hard to read, but it is a fascinating one. It is the grave of seafarer, Captain Tattersell.

Legend goes that one night he overheard some men desperate to get to France. It is October 1651. In the dim shadows of the inn, Captain Tattersell recognised none other than the heavily-disguised son of King Charles (who had been executed two years earlier in 1649). Charles II was on the run after his defeat by Cromwell at the Battle of Worcester. For weeks he had evaded capture. With a huge bounty on his head he had hidden in the famous oak tree, and some say, even dressed as a women (tricky as Charles was over six feet tall), anything to evade capture and certain death.

Captain Tattersell immediately offered his services, free of charge, even though it would mean his public execution for helping Charles.

As soon as the tide turned they left for France and he safely deposited the men on the beaches of Normandy.

Captain Tattersell kept his mouth shut and carried on as if nothing had ever happened, collecting and delivering coal along the coast. Years went by, and after the restoration, Captain Tattersell started to boast in his local, called the Old Ship Hotel, of how he had helped Charles and saved the monarchy. Of course, the locals just laughed at him. However out-of-the-blue recognition for his services came from the Crown and he was awarded a pension of £100 a year for life from the Merry Monarch himself. A vast sum in 1660. With his money, he bought The Old Ship Hotel and spent his last years retelling how he saved Britain. His grave at St Nicholas' states... Captain Tattersell, he preserved the Church, the Crowne and the Nation. How brilliant is that!

I was on my way down to the Whitehawk Estate wondering if my car wheels would be safe. After passing a vehicle with a smashed window, I parked next to a car on blocks. I was already in a nervous state after having to fix a sewing machine in Ditchling next to a couple of Alsatians, both rescued from Ireland. One left me alone, but the other sat by my toolbox, staring me in the eyes, and growling every time I picked up a tool. It was the fastest sewing machine repair I had done in ages, and when I left, I tried to think back if I had even fixed the woman's machine or just panicked and left!

I stood by my Land Rover with the tool box heavy in my hand and my customer's details in a folder tucked under my arm. Well, I had little option now. Whitehawk had gone from bad to worse, and now even the buses stopped running after dark. It was only temporary. Like all estates they go up and down, and many of my customers talked with fondness of the good old days on the estate, which they were sure would return.

The lift was out of action and graffiti surprisingly brightened up the miserable, grey walls of the stairway. There were some surprising details about Miranda, and Chad was apparently crap at rap! Someone had dropped something nasty on the floor, which I gingerly stepped over. I made my way up, knocked at the door and waited. There was a sign on the door that read...'Husband and dog missing...Reward for dog'... I was smiling when a large, happy round woman answered and showed me into the lounge. It was a tidy room with a neat stack of papers in one corner, a dining table and a plasma television that was at least four feet wide stuck half way up the lounge wall. The air was full of cigarette smoke, something that I had grown up with and for some reason always triggers fond memories of my Dad's first puff of the day as he sipped his morning coffee. A vase of fresh pink-and-white gladioli were on the windowsill surrounded by family pictures.

On one of the sofas sat an enormous man, possibly 26 stone, maybe more. He looked naked but as I glanced again through the slits of my scrunched up eyes I saw that he was wearing Y-fronts which were squeezed into oblivion by his hairy barrel-belly. He stubbed out the end of his cigarette. "I know what's wrong with the blighter. The suppressor's blown. Used to do a bit of electrical work in Dubai a while back. Sall won't let me touch her machine. Hey Sall," he shouted, "You won't let me touch your machine will you? She thinks my fingers are too fat to do the job properly. I'm just big boned me, that's all."

"You keep your hands off my little darling," shouted Sally from another room. "White smoke was pouring out of the bobbin winder hole, which was more like a chimney. I expected to turn on the TV and find we had a new Pope! Mum gave me that machine when I was 21 and I want it fixed by a professional, not that lump on the sofa."

"Told you," he laughed. "Don't mind, do you?" He posed, more as a statement than a question, as he held up a fresh cigarette.

"No of course not. I am quite partial to cigarette smoke. You carry on and I'll sort the machine." I sat on the opposite sofa with the machine on a small occasional table in front of me and got to work. He was right, the motor suppressor had overheated. It was an early wax-and-foil type made to suppress electrical static, which caused the old valve radio and television sets to crackle. They were pretty obsolete now and I got down to the business of replacing it. Sally brought me a cup of tea on a tray with a bowl of mixed brown and white sugar lumps and five chocolate digestive biscuits. I hadn't asked for a drink but it was pretty obvious that she knew how to keep a man well-fed and watered.

As I worked, the huge husband decided that he wanted to see more of what was going on. I looked up in horror as I realised that he was going to sit down on my sofa – which suddenly seemed extremely small. The seat under me rose like a tidal wave as he phoomphed himself down and I was pushed towards the padded arm, where I jammed. I was immediately aware of the heat that he was throwing off and tried to avoid his hairy leg as I picked tools from my toolbox by his foot. He smelt of tobacco and Imperial Leather soap, and being almost naked made me rather uncomfortable. He picked up one of my digestive biscuits, examined it carefully as if inspecting a rare object, glanced backwards to where Sally was in the kitchen and then put the whole biscuit in his mouth in one go.

"Told you," he smirked in a self-congratulatory tone. He said it with such force that he projected biscuit crumbs from his mouth over the table. "Hey Sall, here's the offending article," he shouted, holding up the suppressor. "This little sod has caused all the trouble. I expect the motors fine?" he finished looking directly at me.

"Yes the motor will be fine," I said trying not to make eye contact with my new couch buddy as he was so close to me. I started to speed up my work and was aware that I was heating up with anxiety due to being almost glued to the mostly-naked man who was like a radiator himself. In the back of my mind I also had another problem. I had convinced myself that I was not going to get paid. I had already had a bounced cheque that week, two people promising to send me money, and a heated argument with the bank manger, which ended up with him pointing out to me that it did not matter how many guarantee details there were written on the back of the cheque, the woman had no money in her account, and 'HE' wasn't going to pay me!

All that worry was about to be momentarily forgotten when Sally brought out a reel of Sylko thread and placed it on the table in front of me. "Have you ever seen one of those before?" Duh! Of course I had. I had seen many thousands of them. I was more polite, "Yes I have seen a few."

"Not like that one I bet?" Now she had my attention and I stopped work and picked up the reel of thread, expecting to see a very early or rare reel, but alas it was perfectly ordinary. "Well it looks just like any typical reel from the 1930's to me," I said smugly. I loved researching the old companies. I knew well that Sylko had been started by Mr Dewhurst and his sons had eventually taken over the firm. I always loved the way that they never called their colours by their real names. The reels would by no means say pink, they would say 'Coral' or 'Sunrise Pink'. There would be many greens – 'Sea Mist Green', or 'Light Apple Green' or 'Orchard Green'. In later years, Dewhurst's had become English Sewing Limited. Oh yes, I was an expert all right, a cocky one at that, and as we all know there in only one way for experts to go...Down!

"It may look like a normal reel of thread but look at this." She carefully held the reel in one hand and unscrewed the top with another. Now I was riveted as I knew nothing like this had been made by Sylko. I looked across at my fat friend who was smirking, knowing full well that I should never have been so smug with his girl.

From the inside of the reel she plucked a small, square yellowing piece of material. She delicately unfolded it before me. "It's a silk map of German Occupied Territory." "What? Why? How can that be?" I was totally confused. Sally went on to explain.

"My Father brought it back from France after World War Two. He told me that sometime during the early part of the war, word was quietly sent out from the War Office to groups like the Girl Guides and Boy Scouts to call on homes in their area and ask for a donation of an old reel of thread to help towards the war effort. It had to be old, not new, and wooden, the size of an old penny, which most were.

No one was told the real reason for the reels. Thousands of the reels were collected up in old pillow cases and sent to a factory which hollowed them out perfectly, making a hidden compartment inside. The reels were then apparently sent via the Red Cross, to prisoner-of-war camps. Inside there could be folded money or a tiny compass, or like this one, a map. Mixed in with all the normal threads and other items from the deliveries the Germans never found out. It was one of the little secrets of the war. Now that has caught your attention hasn't it?"

"I'm fascinated." I replied, swallowing humble pie as quickly as I could. I made a quick mental note not to sound so bloody smug all the time as it always ended with me looking the idiot.

"I love learning about the sneaky little tricks that we used to get up to." I glanced across at my Siamese twin, momentarily forgetting that I had become glued to him. He was rolling his cigarette between his fingers, as fascinated as I was. "She won't let me sell it," he added as if he had guessed my next question. "It's her only family heirloom."

Sally picked up her precious reel and went back towards the kitchen with a little bounce in her step, obviously as delighted as I was with her thread. I looked down to see that all my biscuits had somehow disappeared, so I got back to work on the sewing machine, trying not to wriggle too much and jam myself even deeper into the sofa.

After what seemed like an age, I ran through a sample piece of cloth, balanced up the tensions and packed up my toolbox. Fat Man examined the fabric, exclaiming to Sally that it was as perfect as he had ever seen it and that there was no excuse now for her to put off repairing the ripped gusset his new trousers. I went to stand and realised that I was wedged into the sofa. I leaned forward and pushed on the sofa arm with little effect. Suddenly he leaned away from me and I almost popped out, stumbling forward a little. He did not look at me, but there was a knowing grin on his face and I had the feeling that he had enjoyed his afternoon's entertainment, more than watching his huge TV.

I was just about to come to the awkward point of asking for payment when Sally appeared out of the kitchen and pushed the cash in my hand. "You did my sister's machine last year and she said that you were a little gem so I knew you would do a good job. Here is a little extra for you as well. You buy yourself a drink on me."

"Well call me a doctor, she's paying cash!" Exclaimed Fat Man, wobbling to his feet in amazement. "She wouldn't let me have the middle of her Polo Mint, and you get cash. I'd hoppit before she snatches it back."

"You cheeky begger," shouted Sally with a sudden fire in her eyes. In one swift movement she stepped forward with her arm raised. Fat Man made a move for the other sofa, but he was way too slow. She placed a perfect slap across his mainly bare backside where his Y-Fronts had ridden up into his butt-cheeks. As he squealed in pain the smack almost instantly left a huge red hand mark. I couldn't help but stare as he sunk into his sofa, then lifted his cheek and sat on one side, screwing his face up in pain. "That will teach you," Sally shouted over her shoulder as she opened the door for me. "He loves it really," she whispered. She winked at me and closed the door.

You just never can tell, I thought, as I counted the wheels on my car. The people who you think would never cheat you – do, and the last person that you would think would never pay up – did, and tipped me as well. I pushed my toolbox into the back of the Land Rover, slammed the large back door and jumped into the car. A group of lads on battered BMX bikes were watching me from the green. I left Whitehawk and rolled down to the seafront and back onto the Eastbourne Road. I had to wonder if they had already checked out the price of my wheels on Ebay!

As I slipped around the Newhaven through road I glanced across at a small tributary of the River Ouse. The tide was out and on the far mud-bank was the faint outline of a boat. It looked like some long lost Viking longship. It seemed like only yesterday that I used to see the old man who lived on that boat. He had a sign on the side saying 'Bosun's Bunk'. He was a retired sailor who spent most of his day watching the traffic go by and smoking his pipe. As the tide went out the boat lay at an angle and he would have a little fire and kettle on the deck where he would knock up a brew. At low tide he would replace his normal stool and kettle holder with specially angled supports that sat straight on the crooked deck.

Years went by, and then one day I noticed that he was not there. The boat which he maintained fell into disrepair. One night, after a storm, it took on so much water that it stopped going up-and-down with the tides; it just stuck in the mud. Slowly the boat rotted away and became a hulk, lying in forgotten silence as the tides washed away more and more. Now just the outline survives, and no one remembers the name of the old sailor who gazed longingly out to sea each day, tapping his pipe on his boots and waving to the kids going by. Such is life – and death.

I stopped at Mac Donald's just over the bridge at Newhaven for a nice Latte, courtesy of my new friend and her blown suppressor, who would always be known to me from now on as Fast-Slap Sally. I could still smell Imperial Leather on me and a nice cup of coffee would perk me up.

Suppressors are a regular part of my work. They are in most electrical appliances from food mixers to drills, and explode at any time causing all sorts of havoc and problems. When most radios and televisions had valves, faulty suppressors caused no end of irritation. Imagine trying to listen to your favourite music with a constant clicking coming out of the speaker as your radio picked up and amplified the interference from your neighbour's lawnmower or some other electrical item. It used to drive people mad.

In 1960's Eastbourne, as I guess was the case in most large towns, there was a specially trained man employed, I believe, by the General Post Office, GPO, who was known locally as the Suppressor Man. I have heard tales about our local suppressor man and one funny incident in particular, which I hope I have remembered roughly right.

If you had constant interference on your television or radio you popped into your local post office and filled in a form. The form was passed to the Suppressor Man who in Eastbourne, for a while, was Mr Chadwick. He was an expert and highly thought of. Sometime later he would call on you with his big box of tricks and track down the culprit which was causing the interference. It could be anything from a badly suppressed food mixer to a faulty washing machine. He would then identify the source with his magic box that picked up the interference sent out by the faulty machine. Once repaired you could turn on your television or radio and enjoy it rather than scream at the thing as it hissed, crackled and clicked.

THE SUPPRESSOR MAN'S SONG.
If in trouble or in doubt,
Cut the little beggar out.
If it causes grief or pain,
Put the beggar back again.

Chadwick was a specialist in a narrow field. He was a busy man and sometimes you would have to wait for a week or more before he could turn up and try to discover the source of the problem that was ruining your viewing or listening pleasure. Chadwick's pet hate was motorcycles, which could not be caught or suppressed and sent his instruments wild as they roared past. His magic box was positioned in front of him, held on by a big strap around his neck.

He was brilliant at his job, and apparently the GPO received countless letters commending Chadwick on his work and excellent service. However there was one area of town where the locals were not so happy with him! In fact, for a time, his name was mud.

There was an unidentified suppressor problem somewhere along the South Street, Grove Road area in the middle of Eastbourne, and try as he could, poor old Chadwick failed to find the source of the problem. Constant letters flooded in and complaints followed. People could not watch television or listen to their music. The area became known as the Bermuda Triangle and something suspicious was going on. Chadwick would get a call and rush down to South

Street only for the signal to disappear. Occasionally he would get a reading, but no sooner than he homed in on the culprit, it stopped!

Weeks went by and try as he may, Chadwick was getting no closer to the source. Was he too old for the job? Was he losing his touch? Did he need replacing? The magician could no longer perform and while he could still track down other sources of interference, he came to a full stop in South Street. Was it deliberate? Was someone watching him and turning off the interference as soon as he got near? So many questions and no answers. One night while trying to track down a particularly bad bout of interference, poor old Chadwick was at his wits end and he decided to soothe his sorrows in the Eagle Pub on the corner of South Street and Gildredge Road. In the pub they all took turns in taking the mickey out of him and consoling him in equal proportions. "You'll get the b****r Chadwick, you mark my words," said a local, patting him firmly on the back. The landlord chipped in from behind the bar where he was cleaning glasses. "That bloody interference is driving me mad. Last Saturday there was so much noise on the radio I thought I had won the pools, but it turned out that Manchester United had lost two-nil. I was fairly fuming. I was thinking of banning you for life!"

"The only thing he'll get is the sack," muttered one of the drinkers. Several of the men in the pub were only in there because the interference on their televisions was so bad that night, they had decided to have a drink instead of trying to watch the box. Just then his machine, which he had plonked on the counter, started to go wild. The pub immediately fell silent. For the first time Chadwick was getting a strong reading right where he was standing. They all looked at each other in bemused silence. Then it stopped!

"Weirdest thing I ever did see!" commented one drinker with a knowing nod before striking up a match and lighting up his pipe with sharp sucks of air.

"Someone's pulling your leg Chaddy!" shouted another from the snug by the fire. Then it started again. The needle swung wildly on his instrument box. By now the regulars had put down their playing cards, drinks and dominoes and were crowded around Chadwick, all staring at the needle in amazement. One second it was registering full blown interference the next nothing. Chadwick scratched his head knowing that people for hundreds of yards in all directions would be banging their TV's and radios and a few cursing his name. "Dammnest thing," announced one.

"It has to be right here," said Chadwick with authority. "It has to be right here. Turn all the electrics off." The pub went dark, with just the light from the fire and the glowing cigarette ends lifting the bar from the shadows. Inside was silence, the only noise coming from the howling wind outside. Chadwick's face was dimly lit by the glowing dial. He was staring hard at his meter as the needle raised and lowered. "Hang on! The meter is going up and down with the wind. That can't be."

But sure enough as they all crammed closer to the meter it was clearly reacting to the wind.

"Everybody outside," shouted Chadwick with regimental authority. They all crowded out of the pub while Chadwick strapped on his machine and followed. This was the most excitement they had at the pub since the darts final. In the dark they all waited, some rubbing their hands in the cold, shocked from the sudden exit from their warm comfy pub, some folding up the collars of their coats against the wind. They huddled around Chadwick as he moved in a semi-circle trying to pick up the strongest signal. Left, right, left a little, right a little, stop. Bewildered they stood in silence, then Chadwick looked up. Above him was the dimly lit Eagle Pub sign. As the wind blew the sign rocked, squeaking back and forth in the night air. Chadwick looked down, then up, then down, then up. "IT'S THE BLOODY SIGN," he shouted. They all cheered.

"Told you Chaddy would find it," someone shouted. They chatted in agreement, patting each other on the back and shaking hands as if it was a joint effort from them all. They then funnelled back into the pub to celebrate.

And so, at last, the hardest suppressor problem in Eastbourne was finally solved; a poor connection on a pub sign that only caused interference when the wind blew. A few suppressors in the right place and the problem would be gone. Eastbourne's Bermuda Triangle was finally tracked down and eliminated and Chadwick, the suppressor man, was the hero of the hour. Letters of commendation soon followed.

CHILLY CHARLIE

Chilly Charlie spanked his dog,
He kicked his cat and whipped the hog.
Chilly Charlie was as cold as ice,
Until the day he met Sally Nice.
Now he sings to one and all,
And has a smile from Spring to Fall.
Chilly Charlie has changed his tune,
Since Sally Nice made him swoon.

THE GHOST OF SIR ARTHUR CONAN DOYLE

"Oh, don't mention that place, it gives me the jitters every time I hear it. A cold shiver runs from the tip of my nose to the end of my toes!"
"Why on earth would that be, Katie? It's such a pretty road running through the golf course."
"It's not the golf course or the manor but what happened to me there over forty years ago that scares me to this day."

But I'm getting ahead of myself. Let me introduce this story properly.

There's a world, an invisible world, of which we know very little. Many shrug and say it doesn't exist. Others laugh at it and at those who believe in it. It's a hidden world of shadows and mystery that sometimes touches us. It's a world that comes to life around the fire on late nights. This is one such tale that was told to me, first hand, many years ago by the person who lived it and it is true.

It all started through a simple accident on my part.

I was at Katie's in Crowborough. I had called on her a few times to service her sewing machine but never really stopped to chat. This time it was to be different. A can of furniture spray would lead to a fascinating tale and the ghost of the great crime fiction writer – Sir Arthur Conan Doyle.

It happened so quickly! One moment I was reaching down to my toolbox, the next I was on my back with a bruised bum. The polished surfaces and profuse amount of mats had been the cause of my abrupt downfall. Above me, on the table, was the sewing machine that Katie had placed on a towel. It hung half-on half-off of the polished table like a boulder on a cliff edge. I held my breath, gently reached up and slid it back onto the table. I fell back onto the floor in agony and heard Katie's pitter-patter of feet rushing towards me.

"What on earth has happened?" she asked as she peered down at me with a worried look.

"Katie, I am so sorry! I slipped off the seat cushion. I went to grab the table but put my hand on the towel. The towel went one way, the seat cushion went the other, and the mat on your hardwood floor shot out from under my feet. It was like trying to balance on marbles. And this is where I ended up."

"My husband used to tell me that I used too much furniture polish; he use to call me Mrs Pledge." She grunted while heaving me up. "I'll get a couple of painkillers and a nice cup of tea for you."

Katie left me feeling a little embarrassed at being helped up by someone at least 30 years older than myself, but I was grateful all the same.

As she toddled off I put my tools back and straightened up the table, paying careful attention to pushing away the carpet and removing the cushion from the flat-bottomed chair. By the time Katie had returned to the room it hardly looked like I had fallen. All that hurt was my pride.

"The day started badly when a car nearly hit me coming by Windlesham Manor," I told Katie in between swallowing a couple of ibuprofen with sweet tea.

"Oh, don't mention that place, it gives me the jitters every time it's mentioned. A cold shiver runs from the tip of my nose to the end of my toes. Sometimes my legs go so weak I have to sit down!"

"Why on earth would that be, Katie? It's such a pretty road running through the golf course."

"It's not the golf course or the manor, but what happened to me there over forty years ago that scares me to this day."

And so one of my favourite ghost stories came to be told because of an accident caused by a can of polish. This is Katie's own story of her encounter with the ghost of Sir Arthur Conan Doyle.

Arthur Conan Doyle, the brilliant writer who brought the super-sleuth Sherlock Holmes to life, had always fascinated me as we have close family ties. He would often stay at the Savoy Hotel while in London, which was owned by Helen D'Oyly Carte, the second wife of my Great, Great Grandfather, Stanley Carr Boulter.

Stanley and Arthur had known each other for some time because Arthur Conan Doyle's uncle, Richard Doyle, had worked with my four-times-great grandfather, James Robinson Planché, on several of his books.

James Robinson Planché, or JP, was the most prolific playwright of the early Victorian era, producing over 170 plays for the theatre and the London stages. He is a forgotten hero of mine and I spend endless hours researching the man without whom I would not exist today.

JP was also an author and renowned expert in historical heraldry, and as Somerset Herald, on call to the Royal Household. He constantly strived to introduce historically accurate costume into British theatre.

In the early 1860's, as JP reached old age, he was commissioned to research a popular old fairy tale. The fairytale had so many origins and variations that there was no definitive version that could be put to the stage or reproduced without argument as to content or copyright protection. JP was no stranger to producing fairytales for the stage. One we all know was *Puss in Boots* which he first staged with Madame Vestris' at her Royal Olympic Theatre, Christmastime 1837.

As one of the pioneers of the Dramatic Copyright Laws, JP knew all about the problems that imitation and pirated plays caused. Also he had been writing fairy tales since his early trip to the Rhineland in the 1820's specifically to collect ancient tales or "Lays" as he called them.

His commission was to construct a classic version of the ancient fairytale to be published for future generations to enjoy. JP's attention to detail, including his knowledge of ancient kings, costumes and heraldry made him the ideal researcher for this important fairytale. JP put his commission very simply…

"I was requested to furnish one solid versification of the many versions of fairytale to accompany the illustrations by Richard Doyle. An old fairytale told anew." J. R. Planché November 1865.

He painstakingly gathered all the necessary information into one definitive tale. There were many accounts of this fairytale. I'm not going to tell you what it was until the end of this little chapter, so hold on a mo. The next bit is a little of the history of our famous tale. I bet you guess it before the end!

The background of the tale fades into the centuries. Charles Perrault published *La Belle au Bois Dormant* in 1697; it was attributed to a story by Italian Giambattista Basile sixty years earlier, *Sun, Moon and Talia*, which was loosely based on another variation of the 1528 romance by Perceforest, which may have had its roots in the Scandinavian story of Brynhild in the Volsunga Saga. There was also a popular poem in the reign of Queen Elizabeth that JP refers to in his book. Jacob and Wilhelm Grimm, the Brothers Grimm, were avid collectors and enhancers of fairy tales, published a version in their collection of stories in 1812.

JP's close friend, Alfred, Lord Tennyson, wrote a poem around 1829 which touched on many of JP's themes including some ancient Greek myths. And finally of course we must not forget to mention Hans Christian Anderson whose charming tales like the *Snow Queen*, *Thimbelina* and *The Little Match Girl* followed the classic fairytales lines of good verses evil.

Now you can see the problem: so many early productions for the stage with varying plots and various endings. No two stage productions of the tale were alike and many arguments followed and no copyright forthcoming. JP was the man to solve all the problems and get a great version for the stage, and many years later, the Big Screen and television. Guessed it yet?

Once JP had gathered all the information he could, he collaborated with Richard Doyle on the illustrations. Richard Doyle was one of the finest illustrators of the day, having worked previously with such Victorian giants such as Charles Dickens and the Brothers Grimm.

Dickens and my distant granddad, JP, were great friends. I remember seeing a copy of a letter from Dickens pleading for JP to pop round and help him with a play that JP had written and which Dickens was putting on for his children. It was a very personal insight into the human side of Mr Dickens.

Anyway, as usual I digress – now back to the fairytale. JP's tale begins:

A Fairy tale should so begin—
A King and Queen
Enthroned were seen,
Their crowns to wear without a care,
One blessing wanted, to their realm an heir

The small book, hardly 50 pages long, was finally completed around 1864 and it would become a classic for all time. It was published the following year jointly in both Planché's and Doyle's names. There were many possibilities for the title because the princess in the story had many names, but no definitive one. To this day it is a quiz question that people will argue about until they go blue in the face. JP was a very careful man when it came to facts and if he could provide no definitive description, he purposely kept it vague. For instance in his book he never gave the princess a name or an exact description of her. It was his clever way to avoid production arguments for the leading lady on stage and allowed any young beauty to play the part. He used the following:

*I leave the reader to decide,
As they may fancy dark or fair;
Whate'er, dears you most adore,
Fancy her that—and something more.*

The possibilities for the title to the book were, *The Beauty Asleep in the Woods, Aurora, The Princess who slept for a century, Briar Rose, Rosebud, The Glass Coffin, The Gypsy King's Daughter, Rosamond, The Evil Mother-in-Law, The*

Young Slave, La Belle au Bois Dormante, and others. I bet you have guessed the title by now! Read on.

The final title for the fairytale that JP chose from a possibility of over two dozen variations was later immortalised by performances in 1890 by Pyotr Tchaikovsky and in 1959 by Walt Disney and Erdman Penner's wonderful version of the tale. Walt Disney used two names for the princess, Aurora, (the dawn) and Briar Rose.

Back in 1856 the fairytale, written completely in rhyme by JP, was simply titled for all time, *The Sleeping Beauty*.

In his version there are seven good fairies or 'fays' and one wicked one, bent and wretched clutching her broom on the desolate moor surrounded only by her crows. When she hears of the birth of the beautiful daughter to the King and Queen she flies into a rage and rushes to the celebrations at the castle.

*In a minute she is mounted and scudding away,
Over hill and dale, over lake and bay,
Over town, over tower, over marsh, over wood,
On that ill wind that blows no good.*

We all know the rest of the story, how the princess is saved after the evil fay curses her, the good fays cast a spell that allows her to sleep rather than die. Years later the handsome young prince forces his way through the over-grown forest to the dormant castle to find his Sleeping Beauty.

I love the way JP ended his tale. After giving thanks in rhyme to Perrault and Tennyson he continues with his tale. The Prince kisses Sleeping Beauty and the spell is finally broken…

She raises to him in sweet surprise,
Her large and loving lustrous eyes.
"It is you, my Prince…" She starts to say,
But who is there to this day,
That really knows what they did say?
Even the first teller of this tale,
Felt there, his information fail.

James Robinson Planché, like his daughter, Matilda Mackarness, was a prolific writer and his version of the old fairytale is one of my favourites of his treasured works. I hope that they would not have objected to me using a small fragment of their work.

Now, once more, back to our ghost story. The family tie with Arthur Conan Doyle had always fascinated me and I had kept an ear to the ground every time I was in Crowborough, in the hope of hearing about the great man and his life. Crowborough is where Arthur Conan Doyle had spent his last 23 years and where he died in 1930.

Coincidentally I had been in communication with Charlotte Crosbie, one of the production team working on Sherlock Holmes 2 with Jude Law and Robert Downey Jnr. One of the scenes (which was later cut), was to be set in a Jewish tailor's shop around 1893. Charlotte needed to know the exact sewing machines that would have been used at that time. I was the man who could help.

It is a constant surprise to me that the town of Crowborough does not capitalise on the huge revenues that a good Arthur Conan Doyle Visitor Centre would generate for them. At present there is a statue of him at the High Street Crossroads, often with a traffic cone on his head, and his home, Windlesham Manor, is a rest home. There is not even a simple museum for one of the most popular authors in English literature.

Arthur Conan Doyle had moved to Crowborough from Undershaw House to escape unwanted publicity. He had visited friends there, staying at the Beacon Hotel on the edge of a beautiful new golf course and fell in love with the area. Crowborough is an ancient town that slowly grew on top of a hill. One of the highest points in my area, the town has its own micro-climate. They say from Saxonbury Hill the next highest place is Russia! I can travel from a sunny Eastbourne up to Crowborough to find it covered in mist or pouring with rain. Originally it was just a high isolated wood where the crows nested, known by the locals as Crow Burgh.

As Sussex folk moved up to 'the place of crows' the name slowly changed to Crowborough. Around 1750 money was left by local ironmaster, Henry Fermore, to build a chapel and school there, for, in his words, "The use of the ignorant and heathenish people of the hill." He must have had personnel problems at his foundry!

However, the school and chapel must have worked, as by the late Victorian times it was a plush town full of large houses with easy access to London and the ports. It was an ideal hideaway for the famous writer. Today Crowborough is mainly hidden by overgrown trees and we only get glimpses of the magnificent views it used to hold.

In Crowborough, Arthur Conan Doyle had his perfect house built, Windlesham. Money had been rolling in from his Sherlock Holmes series published in the *Strand Magazine* and the manor was made to measure with no expense spared. Built on the edge of the green and the golf course, it had a large billiard room and dance floor. It was a state-of-the-art modern building in 1909. It also had sweeping views over the Sussex Weald and down to the sea, views that inspired much of his writing.

Arthur was a popular sight on the golf course and around the town. He regularly walked into Crowborough Woods down to Sweethaws Woods and back up through the Ghyll and home, often stopping at pubs along the way. Arthur used to prefer his woodland walks, as more than once on the streets he was accosted by furious women who hated his outspoken views on the Suffragette Movement.

Another little known family fact is that Arthur collected handcrafted walking sticks, one of which he gave my great, great-grandfather, Stanley Boulter.

Arthur was plagued by family tragedy and became caught up in the solace of spiritualism. His belief in an afterlife gave him a glimmer of hope to hold onto. One fascinating promise Arthur Conan Doyle made was with his son, Kingsley, shortly before Kingsley departed for war, a pact made between father and son that would lead to Katie's ghostly encounter many years later on the golf course.

Arthur made a promise with Kingsley that, should Kingsley die while away fighting, they would meet up each year on the anniversary of his death at their favourite spot, the fourth hole on the golf course. Arthur's worst fear was realised when Kingsley died from pneumonia after being weakened from wounds inflicted during the Battle of the Somme.

Years later, in 1930, the great man himself succumbed to a heart attack and died; he was 71.

One of the most famous writers in history had passed from our world into the next. Just a few days later, at the Albert Hall in London, over 10,000 Christian Spiritualists gathered around an empty chair to try and recall his spirit. They found that although his work may have been immortal, his soul did not respond to their wishes. The chair remained stubbornly empty and 10,000 disgruntled spiritualists left the hall.

Arthur Conan Doyle's body was laid to rest beneath his favourite Copper Beech tree in his garden at Windlesham, just a few feet from his garden gazebo where he often wrote on sunny days. He would say that when the wind was right, he could smell the sea and see the world.

Our story now jumps to October 1956, just a year after the bodies of Arthur Conan Doyle and his second wife, Jean, were dug up from their resting place in the grounds of Windlesham and moved to All Saints' Church at Minstead in the New Forest.

His grave in Minstead is simple and poignant, it states:

STEEL TRUE
BLADE STRAIGHT
ARTHUR CONAN DOYLE
KNIGHT
PATRIOT, PHYSICIAN & MAN OF LETTERS

I remember so clearly coming across his grave by accident whilst staying at a nearby pub in Minstead. In the churchyard I was intrigued by a well-worn path to a simple grave under an ancient yew tree. I followed the path and was amazed to find our great writer resting there.

Using the same technique, I had come across Dick Turpin in York, Spike Milligan in Winchelsea, Isaac Singer in Paignton and Harry H. Corbett in Penhurst. Famous graves always have well worn paths.

The bodies of Arthur and his wife had been exhumed after Windlesham was sold. Arthur would have turned in his grave – if he was still in it! He had loved his garden and the views of the low heather, the green sweeping hills and golden gorse that coloured the far-reaching vistas from his home. It was a place of peace where he had planned to rest for eternity.

The move of the bodies was carried out very quietly and with no publicity or fuss; some say, it was done under the cover of darkness to avoid public outrage.

The whole affair was not what Arthur would have wished for, and shortly after his re-interment is when his ghost started to appear, firstly in his billiard room at Windlesham and then later out in the grounds.

Little did Katie know that she would be one of the people to come face-to-face with him years after he died!

Katie was familiar with Crowborough's best know celebrity. There were many local anecdotes and stories of him. His picture hung in establishments and pubs around Crowborough, most of which claimed his patronage. Katie had heard him described as a distinct, plump large-moustached man of importance who, although not tall of stature, always appeared so.

Although Katie knew exactly what Arthur looked like, with his distinguished pictures all around, she knew nothing about the spiritual pact he had made with his son.

After work, Katie would cut down Sheep Plain towards Hurtis Hill on her way to Jarvis Brook, just below Crowborough. One evening she was making her way home on her bicycle. It was an easy ride, down hill most of the way. On that fateful October night a late harvest moon hung in a cloudless starry sky. Katie felt a sudden chill in the air as she turned toward the golf course and she stopped to tie her scarf tighter around her chin. The trip down Hurtis Hill would be fast and cold.

Little did she know how cold!

As Katie tightened her scarf, she noticed a man wandering around the golf course. Unusual, she thought, at that time of night. Perhaps he was looking for lost golf balls?

Katie gave it no more consideration as she pulled the pedal around with her foot and pushed herself forward on her bike. A low mist was lifting from the warm wet grass into the cool night air of the golf course.

Her eyes had become accustomed to the dark as she started her downhill decent, a full moon and a clear sky meant that Katie could see almost as easily as during the day.

Katie noticed that the man was wandering vaguely in her direction, still looking intently for something. Perhaps he was waiting for someone? She thought as she cycled closer.

She noticed that he was dressed in a very outdated but smart tweed suit. Subconsciously Katie started to pedal quicker. Something about the man seemed vaguely familiar but made her uneasy. He seemed so preoccupied with his search that he never even noticed her. She sped along getting closer and closer to the man. But the mild uncomfortable feeling was turning into apprehension.

Then the man stepped onto the road facing away from her, still silently searching. Katie was now really worried. But she kept cycling, gripping the handle bars as tightly as she could. While pedalling for her life Katie noticed that the mist around his legs was not moving as he walked. Apprehension turned to fear and then to panic!

As she was almost upon him, Katie swerved to the right then left, but the figure wandered straight into her path!

He turned to face her. In an instant she recognised him.

It was Sir Arthur Conan Doyle – dead for over 26 years!

Katie was struck with terror.

But racing downhill toward Sir Arthur, she was unable to help herself. In a single brief moment he looked straight at her. An icy chill shot through her. Katie stared directly into his piercing grey eyes that were full of questions and longing. She screamed and rode straight through him!

She utterly lost control of her bike and collided with the low grass verge on the side of the road. She tumbled onto the grass rolling over and over. Dazed, she lay staring at the stars circling madly above. After what seemed an eternity she managed to struggle to her feet and steady herself. There was no thought of the bike or looking back, she just ran for her life, throwing off her scarf as she scooted down the hill as fast as her legs would go.

Katie was not sure how long it was before she got home but her husband could make neither head-nor-tail of her for an hour. After three large brandies, a period of calming and a plaster on her knee, Katie told her husband what had just transpired back at the golf course.

As much as he insisted, Katie would not go back for her bike or accompany her husband to get it. Her husband walked up the hill for the bike and found it just where she had said, undamaged with a bit of grass stuck in the wheel. He also found her scarf yards away on the Fourth Green, how it managed to get there he could not imagine. He did not hang around on the golf course, but rode the bike home to find Katie snoring soundly by the fire.

The next day Katie made her husband swear never to tell a soul about her encounter with the ghost of Sir Arthur Conan Doyle, and he never did. Mind you, whoever heard the tale would just assume Katie was slightly cuckoo, so it was a wise move on his part.

"About a month later," Katie continued to me over a second cup of tea and a piece of toast, "I bumped into one of my friends, Betty, down at the shops and she asked why I was in such a rush, a few weeks previously, when I passed her on my way home.

"Well, I never even saw her! I saw nothing from the time I got back to my feet at the golf course till the time I got home and was banging on my front door hard enough to wake the dead.

"She said I passed her at 30 miles an hour, running for my life. She wasn't far wrong. I've never seen a ghost before or since and I'll never forget his sad piercing eyes so full of sorrow – never!

"And another thing Alex, I'll never go by that golf course again unless it's feet-first in a hearse!

"I didn't tell Betty the truth. I garbled up some story about being late for supper. I could see she didn't believe a word I said but what else was I going to say? I had just cycled through the ghost of Arthur Conan Doyle up on the green? Oh, and I let him borrow my bike!

"Alex, I have to tell you that you're the first person I have told this story to and I will make you promise the same promise I made my husband take, to tell no one while I live."

It was a promise that I kept. Until Katie's death, I never told a soul about the ghost that haunts the green around Windlesham Manor.

The question that I have never found an answer to is – what was Sir Arthur Conan Doyle looking for? Was it his long lost son on the fourth hole of the golf course, or his own grave that had been moved?

In the invisible world of shadows and mystery we shall probably never know.

KERUNCH

Dark clouds dotted the clear the sky as I drove along the A22 north towards my first call in Newick. The broken sky had allowed a glimpse of the early morning sun and it was rising majestically in my rear view mirror, straight out of a five-star holiday brochure. An area of high pressure had been hanging over the country, sucking in freezing conditions from Siberia. The 'Beast from the East' had kept England colder than Moscow and the gritters had been out each night spreading salt. The dry wind was sweeping the salt across the road; it snaked before me catching the sun, sparkling like a clip from some nature programme on the Arabian Desert. The radio was on in the Land Rover and all was well with the world. A buzzard was sitting on a fence post as I neared Halland. It watched the morning traffic with the same glare that it would watch the final moments of its prey. Perhaps it was waiting for a car to slow down so that it could pounce!

A long, long time ago, before the roundabout at Halland, there was a busy crossroads. Each day on that crossroads stood Maurice Chapman. He was the local AA man. He would stand by his motorcycle and side car saluting motorists as they went by. On his cap, lapels and cuffs were polished brass AA badges that caught the morning light. When his daughter, Maureen, caught the Uckfield bus to work she would wave to him as she passed. He would be saluting the bus and quickly wave at Maureen with his other hand, just a little embarrassed.

I had an appointment with my dentist at 4·30 in the afternoon, so I made a mental note not to dilly-dally. It would mean no stories today and no snacks. I had a sandwich with me to eat on my rounds.

Many people, like me have ordinary lives that roll on, each day much the same as any other. However I had discovered long ago from my customers to bathe in the tiny differences of an ordinary life and to enjoy them. That way each and every day was new and different. It obviously works because I hadn't suffered a bout of 'Monday Morning Blues' since I had escaped the family business 23 years ago.

A flock, or kit, of pigeons were dropping down onto a field where a farmer's truck was parked. In the back of the open truck were bales of wire, posts, and a dirty sheepdog with sharp eyes. It bounced off the truck and raced towards the pigeons scattering them in every direction. The dog circled and came back to its master who was busy repairing fence posts, his breath clearly visible in the freezing air. The dog looked up at him for a moment and then leapt back up in to the truck.

The news on the radio announced that the skeleton found under a Leicester car park was that of the last king of England to be slain in battle, Richard III. I had been sceptical from the start, but DNA testing had proved me wrong and the remains of Richard showed his brutal final moments at Bosworth. I doubt if he had time to shout for a horse, like Shakespeare wrote, as his skeleton, fussed over by forensic specialists, proved that he had been hacked and chopped to death before his naked body was slung over a horse and taken for public display. The last Plantagenet king took to his grave the secret of what really happened to the princes in the tower, but his twisted spine, thought to be a Tudor myth, was actually true! In a final humiliation the skeleton of the young usurper king was on public display and his painfully twisted spine was there for all to see. In death, as on the battle field, his power had deserted him.

The houses along the road were being washed with an orange brush, blazing a line along their bricks and plaster facades. The brilliance of a new day was slowly clawing its way down to their sleepy curtains; below the line the houses still wore their cloak of night, brooding and mysterious. Slowly and beautifully a noise reached my ears from long ago. The radio was playing the song from an opera by Haydn and I was instantly transported back to my childhood.

I am sitting in a large bright classroom at St Bede's private school with 11 other pupils, unlike Ratton Secondary Modern from where I had been ripped, where they had over 40 pupils in many classes and I was happily invisible.

The miners strikes of the 1970's caused power shortages and blackouts in the evenings. The strikes rumbled on until they bumped into Margaret Thatcher in the 1980's. Her iron will broke the men and their families forever. The strikes had triggered mini baby booms. However, unlike servicemen returning to their loved ones after World War Two, our baby booms were triggered by boredom, no heating, no electricity and no television in the evenings. My parents were perfectly placed in the baby business to supply this unexpected boom, and while the miners lost their jobs, my parents soon had some spare cash. I was singled out for boarding school, a trial run to see if it was worth sending the rest of my younger brothers there as well. If they could tame me, they could tame anyone! My lackadaisical attitude to education was my undoing. Without warning, instead of being taken to Ratton, I was taken to the headmaster of St Bede's. I should have realised that something was up when Mumsie offered to drive me to school!

Suddenly I was sitting in a large room with high ceilings in an old Victorian school that was nudged into the foot of the South Downs. Pyemont, the headmaster, was a skinny man living on his nerves and totally devoted to his school. His brilliance on the sports' field, and his ability to delegate decisions, had made a prosperous private school where others were failing. He was also backed up by his beautiful dragon-wife, who had more eyes in the back of her head than Medusa.

So there I was, the stroppy kid thrown into Colditz with the silk hankies. The shock of being taken from the family home and placed in boarding school was only matched by the shock I caused to the class. I made several escape attempts, some temporarily successful, but, much to the amusement of the class, I always ended up back in the joint. Most of the boarders were from other countries, so for them there was little point in escaping.

The kids were physically fit, sports most afternoons toughened them up. However I had a different kind of strength. I was a streetwise yob with five brothers from the toughest comprehensive in the area. I was a rough diamond that the private system was hoping to transform with etiquette and Latin. It was not quite Pygmalion. I learnt, but I also taught, amongst others, an African prince and the son of the chairman of Ford UK, a few new choice phrases from the street. In the dormitories before 'lights out' I would teach the other boarders 'street slang' and the latest songs. My rendition of the Rolf Harris number one in 1969, *Two Little Boys*, went down a storm, and at one point I had taught my whole dorm the words. We would sing it together before one of the teachers would in the end shout at us to keep quiet.

Eventually, with perseverance and a good helping of the slipper I actually settled down and enjoyed my incarceration, though they never quite managed to get the street kid out of me.

It came back to me so clearly, that day in my posh school. The sun was streaming into the classroom as I stared through the large Victorian windows out to the manicured playing fields. Beyond them the silver sea was sparkling like hammered steel fresh from the blacksmith's forge. The music teacher had just picked the boys for the school choir (I'm one), and he opened up an old record player. Just before he played the song he explained a little about it. It was sung in Italian so we would understand little from listening to it, except its emotion. It is a great story for kids and we were enthralled as he told us about Haydn's opera.

The story goes, as I vaguely remember it (I probably have it wrong), is that Prince Ali runs away from his home to avoid his evil and overpowering brother and during his wanderings he spies a beautiful princess. They fall in love and together they make plans and flee, lovers in the moonlight. However they are captured by pirates and sold to separate masters. Ali escapes and spends his time searching for his lost love. He hears that she has been sold to the Sultan of Cairo and is soon to be married to him. He rushes as fast as he can to Cairo searching the streets and markets for his true love.

After explaining the story our teacher carefully placed the needle in the track of the record and played the song. Something magical occurred, angels flew into the room and filled it with the light of heaven. Haydn must have had some of them sitting on his shoulders when he penned it, for suddenly, we were there with him at his first performance of *L'incontro Imrovviso* (*The Sudden Encounter*) in 1775, sitting in the theatre with Archduke Ferdinand. The song, *It Seems To Me a Dream* (*Mi Sembra un Songo*) was a song that touched our souls and filled our bright school room with sound, with love, with life. Blimey, I thought to myself, these posh boys are keeping a secret. At Ratton we were listening to head-banging music from the likes of Black Sabbath, which even now makes me shudder, and now here I was suddenly listening to timeless beauty.

Also, I had been in limbo from being ripped from my family, my old school, my friends and my home. Listening to the magnificence of that song changed the grey and insecure world in which I had been surviving. From that moment on I knew everything at my new school would be alright. My world changed back to colour.

Ali and his princess eventually find each other and escape, but they are betrayed and then captured by the Sultan's guard. The Sultan has the power to have them both executed, however, when he sees their true love, his heart melts and he pardons them both, and they live happily ever after. Well, so our teacher said; it was good enough for us. At the end of the performance we all sat in silence, wishing for the record player to somehow turn itself back on. Except for my new friend Roland Neville that is. Untouched like us, he took the opportunity to squirt ink from his Parker pen over my trouser leg and accompanied it with a horse-bite to my arm, which I suffered in silence. Losing house points because of Roland would be a disgrace. I would get him back later, during break.

After hearing the song I saved every penny I could scrape up, cleaning cricket kits, running errands and helping with homework for the other boarders (that didn't always work out well). As soon as I had enough cash, I shot down to Boots in the high street. Boots the Chemist used to have several floors before the Arndale Centre changed it and on the top floor were books, toys and records. All the latest Matchbox, Dinky, Airfix and Corgi toys would be lined up under spotlights as well as the best selling kids books. Security was tight, and massive black-domed cameras swivelled around on the ceilings. In the record department there were listening booths where you could put your record on and listen to it with headphones before buying.

I bought my record and wore it out listening to it time and time again, until it was so scratched and damaged that it would hardly play. As it started to die I would lick my lips, stick my tongue out and try, as carefully as I could, to gently lower the needle into the right track, and then sit back with my eyes closed. Sometimes the crackling silence would bring the princess to life and sometimes not. I'm sure in my imagination, I was Prince Ali and I was in love with the Princess. In the end the record was misplaced and I forgot all about it until it came onto the radio 40 years later. Listening to it again, after all those years, was like coming across a long-lost friend. My Land Rover has an amazing sound system and I turned up the volume, hearing it as if it were the first time. I glided along with the rush hour traffic, watching a beautiful sunrise being created and listening to Haydn's masterpiece. Life could get no better.

Strangely, although Haydn was the most revered composer of his generation, he was always unsure of himself. He came from humble beginnings as the son of a wheelwright, and rose to become so famous he could hardly walk down the street without being mobbed. He was insecure of his talent and full of self-doubt. For example, after seeing an opera by Mozart he never wrote another opera; he wondered how he could ever compete against such a genius. It was our loss, as one of the secrets of any great work of art is to make it appear undemanding. Artists often work so hard to make something seem so simple.

By the time my trip down memory lane had finished I was almost at Newick, and bang on time. All the cars along the main road as rush hour peeked, were so close together that they could have been connected by tow ropes. I pulled into the slip road opposite the Bull

Inn. Suddenly there was a loud KERUNCH. My car lurched forward and I quickly pulled over to the side of the road. A Renault Meganne had decided to try and drive under my Land Rover. "I am so sorry," came a man's voice across the top of his car. "I took my eyes off the road. Please forgive me. It's so stupid. Oh my god! Look at the mess! Oh no, what is Jill going to say? She is going to kill me."

I looked at his car, his bonnet was badly damaged but his engine was still running, well, clanking. In his hand he had a phone and it became clear what had happened. I looked at my old Land Rover and could not see a mark. We both looked at the back of my car, then the front of his bonnet, then back to my Land Rover. "Bloody French worm steel." I heard him mutter, as it became obvious that this was one-way problem. "Look, you can call the Police if you want but I'll pay for any damage. There is no need to involve the insurance companies."

"I don't think I'll need to claim," I said, with a relieved smile. "In fact I can't see anything at all, except a bit of underseal missing. I can touch that up easily with a tin of spray. That's not going to cost me a fiver."

"Look take twenty," said the man, pushing it in my hand as if it was not open to discussion. He was pale and his face seemed to have drained of blood. A single bead of sweat was starting to run from his receding hairline. His day was not starting well. I could see all the tell-tale marks of stress written over him, not just from this, but his job and family life. It was all going to take its toll in the end, and with his driving, probably sooner rather than later. The next thing I knew he was rattling off down the road, with horrible noises and squeals coming from his car.

Amazingly he was still talking on his phone! I laughed, shook my head, patted my tough old girl and drove to my first customer of the day, only a few minutes late.

The day went like clockwork, and as usual, there was one customer that I could do without. "Back door," said a small head with large glasses that was peering around his front door. He nodded in the direction of the back gate and closed the door in my face. I went around the back to the conservatory. "Shoes off."

It is funny how quickly you can dislike someone but I always persevere and smile as well as I can in the situation.

"I know what's wrong," he said. "It's just that bar – there – knocking. I'm an engineer by trade and that's the problem." I then thought to myself that if he was an engineer why didn't he fix his own machine? He was wrong of course, as most of them are. The problem was just about as far away from the point where he was jabbing his finger as it could get. Still, I settled down with him on one side and his crow-like wife on the other, both giving me instructions.

"It is a lovely machine. How long have you had it?" I enquired, trying to break his monotonous monologue of his amazing engineering abilities.

"Just bought it," said the man, puffing up his little chest like a robin in winter. "Got it for a fiver down the boot sale."

"Five pounds," shouts the wife. "You told me you paid fifty pounds for it! That's just like you, you tight-arse."

I had my head tucked well down inside the base of the machine and no one could see my huge smile. I had to gulp down air to stop myself from laughing out loud. That little twerp had got his comeuppance all right, and from his own dearly beloved. I couldn't wait to get out of the house, as she was brewing up into a right volcano and I didn't want to be around when she went off. As I drove away I thought to myself that it couldn't happen to a nicer chap.

By the afternoon, the day's calls had gone well, and I had been a 'good boy' not stopping to chat, but keeping a keen eye on my watch. I had even popped into a garage and picked up a can of underseal to cover up any scrapes under my old banger. One customer made me smile. On her wall, was stuck a hand-written piece of school paper.

It simply read, *I like the sound my mum makes when she is cross. Sometimes I fink I could throw a bucket of water over her and she would hiss like a fire.* I could just imagine the mother's face when the school teacher handed it to her on parents' evening.

I was poodling along, minding my own business when I passed a sign saying 'Toads Crossing'. I laughed, not expecting any cold-blooded creature to be stupid enough to be out on such a freezing day, but I was soon proved wrong. I saw a car that was coming towards me suddenly grind to a halt, and as-happy-ascould-be, there was a toad bouncing through the traffic trying to commit suicide. I stopped, bunged on my hazard lights, and leapt out. I was amazed how cold the little blighter was as I cupped it in my hands and tossed it in to a stranger's garden. The woman in the car, who had stopped opposite me, gave me a wave as I jumped back in mine and hit the road once more.

My next customer had a sign next to her front door, 'one cat and two servants live here'. She gave me a little shock when I went to find her after I had finished her machine. I had walked into the hallway to find her praying on a small mat. She was kneeling and both her hands were outstretched in front of her. I nearly whelped in surprise. I was a bit embarrassed at having disturbed her prayers. I had seen Muslims praying many times in Egypt and knew not to disrupt their moments of peace. "I am so sorry. I didn't mean to interrupt but your machine is sewing like new."

"Excellent, I had just finished anyway." She rose and rolled up her mat. I tried to break the silence, "So is that way East?" Pointing the way that the mat was laying and knowing that the prayer mat should always face towards Mecca. "Yes," she replied with a curious smile. "Alex, I have an old treadle sewing machine upstairs. Could you tell me if it is worth repairing?"

"Of course, lead the way." The machine was an early Frister & Rossmann that was made in Berlin just before the outbreak of World War One. It had a perished belt and I knelt down to pull it off. From downstairs came a voice.

"Hello Mum. Where are you?"

"I'm upstairs in the spare bedroom...with Alex."

"What on earth?" came her reply. A few seconds later a pretty blonde woman in her late 30's peeked around the corner of the bedroom door. I was still on my knees after unwrapping the belt from under the sewing machine and had the leather in my hand. "Well mother," said her daughter with a wry smile, "this is very Fifty Shades of Grey."

"And your mum's paying me!" I threw in. We all burst out laughing.

Before long I was waiting across the other side of the dining room table as my customer was writing out her cheque. Once again I should have kept my mouth shut. "How many times do you pray each day?"

"I don't. Oh you idiot, Alex!" She roared with laughter. "I have a bad back! I was just stretching it. That's why you made the remark about facing East. Mind you, perhaps I should pray, it may sort my back out."

Ah, I hadn't lost my touch, I thought to myself. If there was somewhere, or somehow, that I could drop myself in it, I always did.

By the afternoon I was making my way home, past Ear Wig Corner, just above Lewes. The allotments looked like some old shanty town outside a South American city. There was a deserted assortment of sheds and chairs, with a few empty bean poles holding a promise of future delights. The traffic came to a halt just before the Cuilfail Tunnel and I sat quietly, slowly edging forward.

On the opposite side of the road was a dead black and white cat that probably died making a dash for home. Its wet body had lost its bouncy fur and it lay sodden and lost in the gutter, waiting for some road sweeper to throw it in his bin. Two smartly dressed boys in blazers and striped ties were walking up the road in the direction of the cat, one pushing a bicycle.

They were laughing, probably talking about one of their teachers or something that had happened that day at school. As they passed the cat the boys stopped and looked aimlessly at it. Quite suddenly the one with the bike dropped it onto the pavement, and grabbed the cat. He held the poor road-washed lump in his arms, clutching it to his chest, as it hung like a rag doll dripping over his smart blazer. His face, that moments before had been a picture of youthful happiness, contorted into that of an old man, every nerve and muscle of his features in a spasm of despair. The tears rolled uncontrollably out of his eye sockets as he stared up and down the street. His brain was confronting the enormity of his loss and his lifeless cat swung with his erratic movements. Suddenly he took off and raced up towards Orchard Road with his dead pet crushed against his heart, leaving his confused friend to pick up his bike. I could feel myself breathing hard and slow. That young boy's early touch with death would be just one of many in his life.

I pulled into the Cuilfail Tunnel and swept into the darkness. My eyes searched for the distant light, but I was dazzled by the headlights popping on as oncoming cars flew into the dark from the other end. I was surrounded by the roar of the road, magnified and encapsulated by the curving earth rather than escaping almost unnoticed into space. The smell of engines wafted into my car and my diesel engine roared back at its opponents like jousting knights at a tournament. Then no sooner was I in the tunnel than I was out of the other side. I rolled into the light, squinting towards the oncoming roundabout. Once over I glided with the traffic around the twisting road cut into the base of Mount Caburn. Before me lay one of the finest sights of our South Downs sprawling across the horizon in a grand sweep of green, as majestic and inspiring as any view in Sussex.

I had the best part of an hour to get to my dentist appointment and my working day was coming to a close. Once more I had experienced a day like no other. It is a fact that no two days are the same, they can never be. Learning to bathe in the tiny differences of an ordinary life, to take the good and the bad, the highs and the lows was something that took years to adapt to but, in that simple fact, probably lies one of the secrets of life itself.

WHO DARES WINS

I was just leaving and was half way out the of my front door, when the phone rang. I went to answer but had that de-ja-vu feeling. I have answered the phone so many times, only to have my whole day interrupted by another 'emergency'. I quickly clicked on the answer phone and listened to the message. A deep foreign voice told me that it was better to park around the back at my eight am call. No problem!

By eight I was wandering up the main road searching for an address that didn't seem to exist. I ran the message through my mind and examined the directions on my customer details. Giving up, I got on the mobile and phoned. The same deep European voice answered that had been on my machine earlier.

"Mr Graham," I said, "I can't find your address!" There was a long pause. "No problem, I shall put white coat on and meet you in road, da?"

I knew I was speaking to the same person from the message left earlier, there was no mistaking that deep accent. Once more I drove up and down the road – nothing!

It was now rush hour and the town traffic was everywhere. Turning around in my car was a real pain. I had passed a woman in a white coat and suddenly thought that perhaps the man had sent his wife out so I pulled up next to her and opened the window.

"Sewing machine?" I shouted.

"Da, that is me," came the deep reply. "You park round back and meet me."

Wow, no wonder she paused earlier when I asked if it was Mr Graham on the phone! It was a woman with the deepest voice I had ever encountered. An eastern European female Barry White!

I parked and followed the girl to the flat. She could have been no more than 25 years old but what an accent, what a voice! On the door to her flat were several signs, no charity bags, no junk mail, no callers. In the flat she started on me. "Here machine, you fix – now," she growled.

Once at the machine, I got to work, stealing sly glances at her for any sudden moves. My hands flew at the speed of light and within a short period the machine was fixed and I sheepishly asked for payment.

"I give you this much" she said, pushing a ten-pound note into my hand.

"Mrs Graham," I said, mustering all my strength. "You know the price that I quoted to you over the phone and this is less than half."

"You not even here one hour, I give you this. Think yourself lucky."

What do you do? Would I dare confront her? Decades of experience had allowed me to fix sewing machines perfectly and promptly – of that there was no argument. I was a quarter Russian myself so you would think that I would know how to handle the situation. I had bargained with just about every race under the sun and surely she could be no harder than my Arab friends. I summoned up all my strength and stared the woman straight in the eyes. Her icy look bore down on me. I went to speak, faltered and finally chickened out. "Okay," I said sheepishly. I took the money and ran to my next customer. What a baby, I thought to myself. On the other hand, at least I wasn't pinned to the floor and made to eat the carpet.

"In 1952 my sister had a new baby, and I bought a new machine to celebrate, to make baby clothes with," my next customer told me, as I dropped oil carefully through the little oil points on her Singer 201. I was so good at it that I could sip tea and drip oil at the same time, possibly one of the best reasons for being ambidextrous. Singer 201's are a rare breed of sewing machine, beautifully engineered and timeless. Even decades later they stitch just as well as the first day that they were made. I asked my customer how she came across such an expensive sewing machine.

"We both went to the Singer Shop. It was in 'Seaside', at the time. I saw this lovely machine, the Singer 201, built up in Scotland. I was told that, even then, that it was possibly the finest sewing machine ever made. The only problem was the price. It was £35 and I was on £5 a week. Between us we only had £28 saved. There was a machine for sale at £28 but I had fallen for the 201. It was so smooth that when it sewed, you just had to smile. You could imagine the Scottish engineers meticulously building every piece and putting it together like a Rolex watch. The salesman told us how the men who worked at the factory in Kilbowie had taken over from their fathers, and their fathers before them. They had passed down their skills for generations in the largest sewing machine factory in Europe. We smiled sweetly at the salesman and he was so nice that he agreed that we could pay the extra when I received my wages at the end on the following month. There was no hire purchase agreement, just a handshake, and we walked out of the shop with this machine. It was so heavy we took turns carrying it home. A month later I returned with the extra money to settle our debt."

"Some years later, my hubby converted it to electric and I have sewn on it almost every week for its whole life. Now I want to leave it to my children and I want it perfect."

"I have to ask, how did it get these scratch marks on the lid? They are so funny."

"Oh," she laughed, "My heart nearly skipped a beat when that happened. I was looking after my granddaughter and I popped upstairs to the loo. When I came down stairs, I nearly died! My granddaughter was sitting on the carpet with one of my knitting needles. In-between her legs was the lid of my beautiful old Singer, and she was scratching it with the knitting needle. I was furious, I went rushing over to tell her off, but when I got to her I couldn't. Instead I started to cry. She had scratched across the lid, 'I LOVE NAN'. How could I tell her off for that?"

I had to smile, kids get away with just about anything sometimes.

I left my customer, headed down to the seafront and along by Eastbourne Pier. It was here a long time ago that one of England's worst peacetime tragedies happened. It all started so calmly. Sam Huggett, the happy-go-lucky fisherman who would always help a friend, was trying to earn an extra crust when his nets came up empty. He illegally took a party of young kids out for a day trip around the pier on his mackerel boat Nancy's Pride. On the morning of Sunday 11 June 1876. Sam broke one of the unwritten rules of the sea: never set sail with thirteen on board.

The harmless fun that started as a sunny ice-cream jolly, turned to disaster as a freak gust of wind caught the boat's fore and mizzen sails, upturning the craft. Samuel Huggett Junior, known to his mates as 'Pork', helplessly watched as one-byone the boys drowned. The full enormity of his actions lay before him and he knew his life was over. He held his arms above his head and slipped down to Davey Jones's Locker to join the children, his salt-washed tears wiped away by the ever moving sea.

People on the beach rushed to their aid but Nancy's Pride had drifted a mile out to sea. By the time rescuers arrived just one boy, Richard Deen, was clinging to an oar. Benjamin Hide reached over his boat and dragged the only survivor to safety.

The rest of the children who were not so lucky, had their limp bodies gathered up and taken to the Victoria Tavern in Tower Place to be cuddled by their heart-broken mothers who, clutching their soaking bodies, rocked in rhythm with the waves that had taken them, praying for miracles that would never come. A nationwide fund was started for the bereaved families, but what price do you put on the life of your child?

At my next call I had knocked on the door and waited, but there was no answer. I knocked again. I listened for any sign 'of movement and then I heard, "Come in luv', no need to knock the door off its hinges." I walked along the corridor of the house and went into the front room where Lucy was sitting in a large armchair. She had a glass of wine in one hand, the *Daily Telegraph* open across her knees, and her feet soaking in a vibrating tub of warm water. I had to smile at the scene.

"Oh it's you Alex. Don't you just love this time of day?" she asked. "That first part of the afternoon when the world slips into a comfortable silence. When you can sit and gather your thoughts before everything starts again."

"I know what you mean. The amount of customers I meet between two and three in the afternoon who are just relaxing, enjoying the moment, mainly before the kids get home."

"I thought you were the insurance man," she added. "I am expecting him as well. My son helped me with my decorating, and while he was using his heat gun to strip off the paint, he set my wall alight which travelled up to the bedroom ceiling. Not a great job. I won't be asking for his help again."

"Maybe that was his plan all along!" I said, smiling.

"Wouldn't surprise me." She said sitting up and laughing, folding her paper away. "He is tricky, but if you saw the look of terror on his face when he was rushing around with my mopbucket, I doubt it. He is a silly sausage but he is my only son and he has a heart of gold, which is why he was trying to help me in the first place." As she finished, there was a knock on the door and in came the insurance man. I fixed the machine and hit the road, wondering how she would show him how the fire had started while sitting with her feet in the vibrating tub.

Have you ever wondered what you would drag out of your house if it was on fire? The most precious, bought or sentimental object, that you might risk your life for? I once met a customer who told me, as her mother had done exactly that. Her mum, Annie Pratt passed down the tale of her old Singer and how she saved it from the Germans.

Annie bought a brand new Singer 27k in 1926 to make a christening gown for her daughter. Because Singers were so expensive she paid for the machine in monthly instalments. Each month she would take her special Singer Hire Purchase book to her local Singer shop. They would put a stamp in it, and then sign it after each payment. This went on month after month, year after year, something that we could hardly comprehend today, when new sewing machines cost so little. She paid monthly instalments from 1926 until 1941 for her Singer sewing machine!

By 1941 the family were living in Alexandra Road in Plymouth. Annie was on the last month of her payments for her machine and was so excited to be finishing of her debt. It was a well known fact that if you stopped, or missed any payments to Singer's they could legally take the machine back, even if you had paid for many years. 15 years of payments were finally over but something quite unexpected was about to happen that changed everything.

Plymouth, on the southern coast of England, was, during the Second World War, a vital naval base. In 1941 Britain was having a bad time with German bombers dropping bombs everywhere. The Blitz was not just on London but covered many of the industrial towns as well. Plymouth was a prime target because the naval base was supplying shipping and munitions for the battles raging in the Atlantic. It was of high strategic importance to both sides and during 1941 the bombing raids on Plymouth became more and more severe.

One day while Annie was at work, the air raid sirens sounded. All the workers made their way to their designated air-raid shelters but, as Annie was in her orderly procession she spotted German bombers dropping loads of incendiary bombs. To her horror they were heading straight for her house. Suddenly Annie broke from the queue and ran for home. In her house was her beloved Singer with just one payment left and she was NOT going to lose it.

As she got closer to her home the bombers circled. They were under furious Ack-ack attack from the heavily armed ground forces around the naval base. One of the bombers headed straight towards her, dropping its load of incendiaries as it roared overhead. She dived under an apple tree along her road, the bombs landing all around her. She looked up to see the houses all along her street on fire. The bombs had spread burning oil over the rooftops.

Annie ran like fury with just one thought on her mind. She raced along the road straight to her burning home and to her most precious item, her Singer.

Along with the windows, the front door was blown out and she ran straight into her living room to find her Singer. She grabbed it, and the piece of new material she had just used her coupons on, and ran out of the house. The arm of her coat was smoking but she made it out to the street clutching her machine.

Luckily, she only had minor burns on her hands. Her family were re-housed and later that month she went down to the Singer shop and proudly made the last payment on her Singer sewing machine. It was all hers at last.

Annie survived the war and she carried on making clothes on her sewing machine. She often told the tale of how she and her sewing machine survived, and how all those years ago, when the air-raid sirens sounded in the April of 1941 she dared to run to her house instead of cover. She said that she only had one thought on her mind, "Those b***** Germans ain't getting my sewing machine!"

EPILOGUE

I am in a steaming bath. I have my earplugs in and the painkillers I took an hour earlier are working. I am floating with my head just high enough above the bubbles so as not to breathe in any water. The pain from my old aching body is easing away. All the accidents of my life that have left their constant reminders are slowly disappearing. In the silence I only have the tinnitus squealing in my left ear. There is no escape from it, all day and night. I had learnt from one of my customers that you have to make friends with your tinnitus otherwise it will drive you mad.

My heart beats slowly and rhythmically, pounding beneath the water in my chest. Since Rob Wicks, my brilliant young doctor, had altered my daily dose of pills, my heart had been as regular as a Swiss watch. Another plus meant no more heart restarts at the Local District General Hospital. The last time I was rushed in when my heart had decided to go into a unique rhythm all of its own, a nurse was trying to get me to agree to sign one of my books. That was in-between me setting off all the alarms on my monitors and being wheeled into emergency. One of the huge benefits from visiting so many older women is that they have endless survival tips. "Old age ain't for the weak-hearted," they would say, "you take what your doctor prescribes and you adapt."

It had been a long day, one of countless thousands that I had worked on my personal journey along the road of life. I lay comfortably in the warm water, like a baby in a womb, thinking how to finish my book. It has been such a pleasure to write and soon it would be over. I remember so clearly all the experts warning me not to bother writing my first book, *Patches of Heaven*. I would never get it published, it would never sell. In the end, I sent my amateur effort off to a printer along with all my spelling mistakes. They converted it to their computer programme and added hundreds more. The result was 1,200 books printed full of awful mistakes. Like the printer, I nearly died when I saw what had happened. However something astonishing blew me away, as within a few weeks I had sold every copy. People were forgiving about the grammar, they just loved the stories. The tales made them feel good and the reprinted versions improved each time.

People started writing to me and even turning up at my door. They came from all over the world, Canada, America, Australia. Then a professor from Mexico turned up on my doorstep with a cake! Jaque Johnson explained that it was all about the content, not the errors. The comments came flooding in, all before the instant age of Twitter and the like. They came on postcards and letters, flying through my door. One day I had a special delivery van with two sacks of mail. I was astounded. My favourite letter was from a lady dying of cancer, she said that while she was not convinced that our world was safe or sane, the edge of it was softened by my words. I nearly cried. And so the books went, one, two, three, and more.

My eighth book, like my life, had flown by. They take years to write, slowly picking up all the stories, researching them to make sure that they are as accurate as possible, and then building them bit by bit, all between my working day.

No fast easy fiction for me. When I handed over *Sussex Born And Bred* to my American publishers it was like giving a child away. I am always thinking that I can add more, I can do it better, I can always do better. With this new book I was delighted to be taken up by an English publisher. My American Publishers, Fireship Press, who had printed the last two of my earlier books, *Sussex Born & Bred*, and *Corner of the Kingdom*, were great but too far away. I couldn't hop on a plane to some American Women's Institute to give a talk and book signing before supper. So Country Books in Derbyshire would be taking my latest work.

I always try to be accurate. I remembered the old dear who had told me that her husband had been a rear-gunner in a Lancaster Bomber. I was entranced. Later he walked through the door. I had assumed that he was long dead, but in fact he had just gone to get his morning paper. I commented on how scary it must have been in the bomber. To my surprise he told me that he was a mechanic on the ground crew. He had never flown in one in his life but he had spent all day changing the spark plugs! I was amazed how his wife had promoted his status from mechanic to flight-crew. If he had not been around, I would have believed every word she had told me.

I have so many unfinished stories. Piles of notes litter my study and they are all calling to me to put them down in print. Over 100 stories waiting to be written before it is too late. Luvina, was supposed to be christened Lavinia, but her dad celebrated a little too much and spelt her name wrong when he went to register her birth. As a 22 year old she came down from Aberdeen and found work at a huge country manor. Armed forces were stationed nearby and used the manor's stable facilities to wash in. She was working there when she received the news that her brother had gone down with his ship. Then there was Arthur, who risked his life riding to and from the Front as a dispatch rider during the Second World War, and Joe who, part of a bomber crew, bailed out, and as per his training, quickly buried his parachute in the snow. He made his way down to some lights and hid in the local village but was soon found by Germans. When he asked them how they had found him so quickly one said in broken English, "Ve followed ze footsteps." His story is astonishing and finishes at Beachy Head with the new Bomber Command Memorial in 2013.

I still have not written about Dan, who as a child, witnessed his local margarine factory exploding, and rushed along with everyone else in the vicinity with buckets, wheelbarrows and containers to shovel up the millions of peanuts that had dropped from the sky like manna from heaven. There is the great story of Jane, conceived on the back of a Triumph Bonneville in a lay-by on the Heathfield Road. Her mum had a burnt leg (from the exhaust pipe), and a baby as the telltale signs of that late night encounter. I have a sad story about the Titanic and two lost sisters, tales of local smugglers and ghosts, loads of ghosts. The touching story of the umbrella man with his licensed pitch on the corner of Oxford Street in London, where he would sell his wares made by his family. His devoted dog would see him off at the bus stop every morning and be there waiting for him every day at the right time, until he never came home. Then there was the extraordinary tale from my old neighbour Mr Flint, a highly intelligent headmaster who, in the middle of a tank battle, floated out of his tank and sat on a cloud and watched from above as the battle raged below him.

It is the same old problem, so many stories, so little time. Whoever invented the 24 hour day missed a trick – they should have made at least 36 hours in a day so that we could fit more stuff in.

I hate coming to an end of a book. I deliberately slow up, which does not go down too well with my publishers. I feel like I am losing my baby. My book has grown up and is leaving me, but each day new stories pour in that I want to include. Today, for example, I was working in Westham when news came rushing in with the local milkman that the rector of St Nicholas' Church had suddenly died. Reverend Dr. Anthony Christian was apparently a distant relation of Fletcher Christian and had been one of the village characters. I am sure that he once kept noisy geese, much to the annoyance of his neighbours. He was a popular and sociable vicar and he loved a tipple at the local pub, the Royal Oak and Castle, opposite Pevensey Castle, and perfectly placed right next to his church. At my sister-in-law's wedding, the only bright sunny day in a month of rain, he had been the first to get from the church across to the pub. He was happily leaning on the bar half-way thorough his first drink when we arrived. We all celebrated together. We realised that wedding days are good days for vicars and a real perk of the job.

On the other side of Pevensey Castle is the church of St Mary The Virgin, the first Norman-built church in England, finished in 1080. The two churches sandwich the castle, and in various sieges arrows were shot from the roof of St Mary's into the castle. When the last vicar retired from the popular Anglican church they threw open the door to the bell tower in celebration and rang the bells. He had stopped the bell ringers getting into the belfry and allegedly changed the locks. Apparently some locals had threatened to dip him in the village pond till he came to his senses. It made little difference, even when mentioned in Parliament. The doors were tightly closed until he retired. Apparently the bell ringers were going to ring a peel of bells to a new tune, *'ding dong the vicar's gone'*. I love this sort of history and it is happening everywhere, even as I write.

Suddenly there was a bang on the bathroom door; it was Yana! "Get out of the bath, you have been in there for ages and now there is a team from Greenpeace at the front door. They have heard that there is a whale stuck in my bath and they want to roll you back into the sea at high tide!"

And so here we are, at the last page once again. It is always the hardest to let go but if I am spared a few more years, I will continue with my tales. I do hope that you have enjoyed coming with me on my travels. I already have a folder on my computer marked Book Nine. I also know, that within weeks I will come rushing home with some amazing anecdote or story from one of my customers that I just have to add to the pile already waiting. Until then my friends, I bid you once more, farewell.

TALES FROM THE COAST by Alex Askaroff

Tales from the Coast, crammed with original photos, continues the series of short true stories in which Alex Askaroff once again brings both England's history and her people vividly to life. Although they are local stories Alex's popularity and easy writing style means his books are now available in over 40 countries. They were some of the first books ever on digital media like the Apple iPad and Kindle.

Alex Askaroff, a Sussex lad, left a thriving family business in specialty textiles to become a journeyman, repairing sewing machines, carrying on a trade that he had known since a child. Alex, now a Master Craftsman, has the enthusiasm of a poet and a pure loveof story-telling. As Alex brings sewing machines back to life he also picks up local stories, history and gossip. And what stories they are! All the stories are inspired by the people who actually lived them. These are real people, no media sensations, just ordinary hard working people who, through their long lives, have had fascinating incidents indelibly burned into their memories.

Tales from the Coast celebrates the spirit of Sussex life, its people, colour and vibrancy. Dorothy tells of her years of hop picking even as the Battle of Britain rages overhead, Sheila tells of her encounter with a jaguar in the jungle far from home. Flo tells of her evacuation as a child and her glorious years on the farm, far away from harm.

From the disappearance of Lord Lucan in Uckfield to the Buxted Witch, from William Duke of Normandy, to Queen Elizabeth's dressmaker, Tales from the Coast is crammed with a fascinating mix of true stories that will have you entranced from start to finish.

Pearly King & Queen Margaret & Brian

Miss Blackpool 1948

You may feel that you know Sussex but I guarantee that Alex will provoke you
into wanting to look again. FC. OBE

Vibrant stories told with skill and humour that provide a valuable insight
into a magical part of England. Alaric Bond. Author

Wonderful humour and writing. Sussex Life

A polished masterpiece.
What's On Magazine.

I couldn't put it down, Hillarie Belloc would have been happy to put his name
to a book like this –
Magnet Magazine

Landscapes and pictures straight from the heart. A fascinating read –

Jim Flegg Country ways Television

Alex Askaroff at Birling Gap.

VISIT SEWALOT.COM

Printed in Great Britain
by Amazon